Implementing 802.11, 802.16, and 802.20 Wireless Networks
Planning, Troubleshooting and Operations

by Ron Olexa

ELSEVIER

AMSTERDAM • BOSTON • HEIDELBERG • LONDON
NEW YORK • OXFORD • PARIS • SAN DIEGO
SAN FRANCISCO • SINGAPORE • SYDNEY • TOKYO

Newnes is an imprint of Elsevier

Newnes

Newnes is an imprint of Elsevier
200 Wheeler Road, Burlington, MA 01803, USA
Linacre House, Jordan Hill, Oxford OX2 8DP, UK

 Recognizing the importance of preserving what has been written, Elsevier prints its books on acid-free paper whenever possible.

Library of Congress Cataloging-in-Publication Data

(Application submitted.)

British Library Cataloguing-in-Publication Data
A catalogue record for this book is available from the British Library.

ISBN: 0-7506-7808-9

For information on all Newnes publications
visit our website at www.newnespress.com

04 05 06 07 08 09 10 9 8 7 6 5 4 3 2 1

Printed in the United States of America.

Contents

Preface

Communication, a word that many associate with modern technology, actually has nothing to do with technology. At its core, communication involves nothing more than the spoken or written word, and symbolic languages like art and music. Technology has become synonymous with communication because technology has historically been the method by which communication to or by the general population takes place. From the printing press to the telephone to radio and TV broadcasting, technology has touched our lives by providing convenient ways for a large population to communicate. Because the intertwining of technology and communication is fundamental to our culture, the technology of communication in a way defines our culture.

As we enter the early years of the 21st century, humanity is awash in instant communication based upon the radio technology that makes it possible. As a society we have near real time access to world events occurring in any corner of the globe, and as individuals we have instant voice communication with each other by virtue of the telephone in its many shapes and forms.

Over the last decade of the 20th century, the cell phone redefined our cultural expectations of communication. The advent of the portable phone along with price competition among hardware and service providers has brought true personal voice communication service to a large segment of humanity.

But what of symbolic communication? Computing power and flexibility have allowed us to digitize these communications to make their dissemination more convenient, and the portable phone has set an expectation that all communication requires portability, convenience, and cost effectivity. Unfortunately, these voice-centric systems have only marginally addressed the more complex nature of symbolic communications. Moreover, as a society we are no longer content with simplistic communication. The advent and adoption of the computer and the myriad software packages available for it has offered the ability to generate a new wave of symbolic communication combining art, pictures, music, and words into a targeted multimedia presentation. No longer is a generic presentation enough. It is now so easy to tailor the presentation of information to an individual or group that audience-tailored and targeted multimedia presentations are now expected.

By their nature, these presentations are large and require high bandwidth transmission facilities to accommodate their rapid dissemination. Such facilities are available within a wired office LAN or to some extent in the wired telephony network (and by extension the Internet), but these facilities only serve a segmented local community. The user who is not connected to a wired broadband facility cannot gain access to this communication.

While there are several low speed portable data systems operating today, their speed makes them useful for only the most rudimentary of communication: short written messages like email, or small low resolution images.

The growing volume of targeted multimedia presentation material requires a high bandwidth delivery facility. Couple this with our society's need for mobility, and you quickly realize that currently available ubiquitous coverage wireless data delivery solutions fall far short of the bandwidth required by this emerging communication requirement.

Enter Wi-Fi (Wireless Fidelity). Late 2002 through 2003 has seen a remarkable interest in 802.11 (a.k.a.: Wi-Fi) network deployment. The 802.11 standard is a wireless Ethernet standard that was designed to simplify office LAN deployment by eliminating wiring requirements. Interestingly, with the advent of Broadband Internet connectivity in the home, this technology has found its niche not in the office, but in the home. Wi-Fi capability can be added to a computer or PDA by simply plugging in a card. Laptop computers now come with Wi-Fi functionality preinstalled, and consumer quality Wi-Fi base stations cost less than $100, and are becoming easy to install. Wi-Fi has become the equivalent of a cordless phone for your computer, and just as the advent of the cordless phone presaged the development and consumer acceptance of cellular and PCS services, the adoption of Wi-Fi may be giving us the early glimpse of the needs and expectations of the next generation of wireless data consumers.

The deployment of Wi-Fi and other wireless data delivery technologies have not stopped at the home and the office. The fact that these devices operate in unlicensed spectrum is allowing individuals and companies to take the next steps in deploying area wide wireless data networks. Today more and more systems are being deployed to provide public Internet access in public areas as small as a coffee shop (the "hotspot model") or as large as a community (the "WISP" model). Some companies are implementing these systems in order to provide "for sale" service, while other networks are being implemented by individuals, companies, or groups to offer free Internet access to those who enter the covered area.

By late 2003, the first 802.16-based equipment began to enter the marketplace. The 802.16 standard is designed as a next generation broadband data delivery system for Metropolitan Area Networks (MANs). 802.16 overcomes many of the shortcomings of 802.11 when used in a MAN environment, and can operate in licensed and unlicensed bands from 2 GHz to 60 GHz. The additional spectrum, bandwidth and throughput capability of 802.16 will markedly improve wireless data delivery, and should allow even more wireless data service areas to be deployed economically.

The initial 802.16 specification effectively offers a solution for providing high speed wireless communication to fixed locations, but it still does not offer true mobility. The 802.16e and 802.20 standards are being developed as next generation solutions that can offer true high speed (over 2 Mbps) connectivity to vehicles traveling at over 90 MPH.

The successful development and growth of any high-speed wireless data network, be it standards-based or a proprietary solution, requires not only an understanding of the data network requirements of the system but also, maybe more importantly, an understanding of RF propagation and interference management. There is a significant volume of published information that describes the fundamentals of wireless LANs and 802.11 technology and/or the networking needs of such a system, but little information is available regarding the radio propagation issues surrounding the deployment and management of any of these RF-based technologies.

This book is intended to correct this information gap by providing a basic understanding of the issues surrounding the implementation of RF-based technologies. It will be useful to anyone who plans to implement a wireless network. The concepts in the book are equally valid for systems using licensed or unlicensed spectrum, and apply equally to any technology selected for implementing the wireless network. It is written for an audience that has limited or no RF experience, and will offer the reader a basic but practical understanding of the concepts behind RF, such as transmitting and receiving, antennas and their effect and use, RF propagation characteristics, interference management, and regulatory issues.

These concepts will provide the underpinnings for the later chapters, which will focus on the real-world issues of designing, implementing, and optimizing a wireless network.

This book will provide a solid general understanding of the issues encountered in designing and deploying a wireless network, and should help you gain the ability to effectively plan and construct a network either by yourself or with the help of others.

The primary audience for this book is the technical professional responsible for deploying a wireless data technology. This audience can be as diverse as the IT professional who now has to cope with adding wireless connectivity to an office LAN; the manufacturer, or Value Added Reseller (VAR) who is selling wireless hardware solutions; the individual, group or community that wishes to learn enough to be able to effectively deploy this technology to improve the services available to an area or community, or the engineering professionals that will eventually lead the development of the ubiquitous carrier class wireless data services that will someday become to data what cellular and PCS have become to voice.

The secondary audience is the management, sales, marketing, planning, or accounting professional and the investment community that supports development of new technologies. The content of this book will be valuable to anyone who is responsible for the decision to adopt these new wireless technologies as part of a business plan, and is now facing the problem of not having the appropriate level of knowledge to "ask the right questions."

It is my hope that this book will educate and inspire a new generation of wireless pioneers, and in doing so assure a bright future for wireless data services. I look forward to the day when our cell phone evolves into a pocket-sized device that will fulfill all of our communication needs, and allow our words and visions to instantly reach the farthest corners of the globe.

Introduction

For nearly a century, voice communications networks have been king of the hill. The growth of the Internet in the 1990's has given rise to the need for data communications networks that are as well developed and flexible as the voice networks that are currently in place. Broadband connectivity has become available in certain places using the wired infrastructure of Telephone and CATV companies, but the coverage provided by these systems is far from ubiquitous. While these systems can address the low speed data requirements of many individual users, the high speed (>2 Mbps) data communications requirements of corporate entities cannot be easily and economically met using copper wires. Fiber optics can and does offer the possibility of enormous speeds, but operational fiber cable to the home or business is not commonly available, and will not be for the foreseeable future.

In addition, the need for instant data is becoming as urgent as the need for instant voice communications. Instant voice communications needs are met by cellular-like services. Although these same providers have begun to offer data services, just like their wired counterparts, their speeds are constrained by the limitations inherent in systems that were originally designed for voice communications.

Wireless-based data networks have the ability to meet the data requirements of corporate networks as well as the broadband fixed and mobile requirements of individuals. There are numerous spectral allocations that can be used to provide these services, and while some are licensed, thus limiting their deployment to those who have access to the spectrum, others are not. It is in this unlicensed spectrum where many wireless data needs are being met.

The year 2003 has seen a remarkable interest in wireless data network deployment using the 802.11b standard. Most intriguing is the fact that many deployments have nothing to do with the service originally envisaged by the developers of this technology. The 802.11 standard is a wireless Ethernet standard that was designed to simplify office LAN deployment by eliminating wiring requirements.

However, today more and more systems are being deployed to provide public Internet access in areas as small as a coffee shop (the "hotspot model"), as large as a community (the Wireless ISP or "WISP" model), or anything in between. Companies are implementing some of these systems in order to provide saleable service, while other networks are being implemented by individuals or groups with the intent to offer free Internet access to those who enter the covered area. As 2003 comes to a close, the next generation of hardware operating under the 802.16 standard is beginning to enter the marketplace. Wireless ISPs are evaluating this technology as a better solution for metropolitan area network deployment.

The surging interest in deploying wireless technology to provide public Internet access is made possible by several factors:

First and foremost is an unmet need. The Internet has become as much a part of our lives as the telephone. Unfortunately, broadband access is still not a common commodity. As the user's requirement for constant data connectivity grows, so does the expectation that connectivity will be available when it's needed, whether that's at home, at the office, or at an airport, train station, or coffee shop. Where such connectivity is not available, we have a groundswell of individuals, groups, and corporations trying to "fill the gap" and provide connectivity wherever they believe the user will want it.

Assisting these entities is the second contributing factor: free spectrum is available. The FCC has allocated spectrum in the 900, 2,400 and 5,600 MHz bands to Part 15 devices. That's good news and bad news. The good news is that the spectrum is freely available to anyone at no cost. The bad news is that the spectrum is freely available to anyone at no cost! The Part 15 allocations are shared by many users and thus, to limit their coverage and interference potential, carry significant limitations on power output. Additionally, portions of these bands are also allocated to licensed users like Amateur Radio operators, Government, and Defense users who have "primary" rights to the band. This means that these licensed users have overriding rights to the band. If a Part 15 user causes harmful interference to a licensed user, the Part 15 user must either correct the interference or cease operation. On the other hand, Part 15 users must tolerate any interference from any source with no recourse to correct the situation. In other words, unlicensed users in these bands are subject to uncontrolled interference from any other users in the band (including other Part 15 users). This leads to some risk when the unlicensed bands are used as the basis for a business. Power limits and interference will limit the distance over which part 15 devices can effectively communicate. These issues and will be examined in detail in the planning and implementation chapters of the book.

In a move that may presage a change in policy on these bands, on Sept 17, 2003 the FCC released a Notice of Proposed Rulemaking (NPRM) that increases the flexibility of using Part 15 devices in a community coverage or Wireless ISP (WISP) environment. Under the rules proposed in the NPRM, the power limits associated with equipment in these services will be allowed to increase significantly when directional or "smart" antennas are used. This development could lead to cost effective coverage of large areas using publicly available Part 15 spectrum. Of course, it's still shared spectrum, and interference must still be managed, a job potentially made more difficult by the higher power authorizations.

The third factor leading to widespread deployment is standards. The IEEE and the Wi-Fi alliance working with and through industry promulgated a standard that let any manufacturer produce a device that would work with any 802.11x equipment. This level of standardization makes implementation easy, as any client card should be able to communicate with any other 802.11x hardware, regardless of manufacturer. Similarly, device compatibility for the 802.16 standard is being promoted by an organization called the *WiMAX Forum*. This compatibility assures that any user's hardware will work on any system deployed using the appropriate standard. In addition, this standardization helps to drive down the cost of hardware.

The fourth factor is cost. Because of freely available spectrum and the 802.11x standards, inexpensive equipment exists. And it will all inter-operate. Since there is no requirement for licensing, this technology is as easy to acquire as a cordless home phone (that also, by the way, probably operates in the same band). This allows manufacturers to ramp up production and gain the benefit of the economy of scale associated with producing tens of millions of devices.

The reduction in cost drove many users to begin installing 802.11x hardware in their homes and offices, thus leading to a significant deployed equipment base.

Now many of those users are discovering the benefits of mobile computing, but their usage of this emerging service is still limited to where service is available.

This brings us full circle. Free spectrum and equipment standards generate inexpensive equipment. Inexpensive equipment generates consumer use. Consumer use generates expectation. And expectation generates business opportunities predicated on filling these unmet needs, using the same free spectrum and cheap hardware that has allowed 802.11x to develop to its current state.

In order for these new entrepreneurs to begin to build more and more complex networks using this (or any other) technology, they will need to understand more than the networking requirements of a data network. They will also be compelled to gain a solid understanding of the characteristics of the RF system they intend to deploy, and the effects of environment on that RF system. Without an understanding of the RF system and RF related environmental challenges, the deployed system may be more costly than necessary (and you may never realize that you are wasting money). Worse yet, it may never work as expected and you'll have no idea why.

The later chapters of the book will focus on the real-world issues of designing, implementing, testing, and optimizing a wireless network. These issues will include how to select equipment most appropriate to the service being contemplated, as well as understanding the cost variables associated with both the capital expenses (CAPEX) and operating expenses (OPEX) associated with a particular solution.

This book is divided into chapters, each dealing with a specific aspect of radio, RF, or system design and deployment. The content is organized in such a way as to develop fundamental understanding of a subject that can be built upon in later chapters. While relatively few chapters depend directly on a previous chapter, most are dependent in a general way on the information presented in earlier text. Those with prior RF experience are welcome to skip around, but I suggest that most readers follow the normal reading sequence.

Later chapters will provide guidance for planning, designing and implementing a wireless network. Both point-to-point and point-to-multipoint networks will be discussed. Three example networks will be featured: a hotspot, a large office

LAN, and a WISP network. Additionally, there will be a discussion of the design requirements of a fully mobile network. The unique planning and implementation requirements of each network will be identified and discussed. These example systems are designed to provide coverage of a small geographic area. As the need and expectation of universal ubiquitous data service grows out of these small scale networks, larger networks will be developed. The 802.20 standard is designed to bring mobility to broadband data services. By definition, mobile services have ubiquitous coverage. Among other tools, the CD-ROM accompanying this book contains a spreadsheet that can be used along with the radio knowledge gained from this book, to determine CAPEX and OPEX estimates for designing and constructing such a system in most U.S. markets.

While the examples in this book mostly focus on the use of unlicensed Part 15 spectrum, the principles, tools and techniques can be applied to any RF-based network. Ultimately if you are successful in deploying a network based upon unlicensed devices, you will have no problem using the knowledge and experience gained to deploy a better system using licensed spectrum.

What's on the CD-ROM?

The CD-ROM contains documents and tools, which can be useful in designing a wireless network.

- A fully searchable eBook version of the text in Adobe PDF format.

- Calculation Spreadsheets

  Channel noise floor and minimum signal calculator is a spreadsheet that can be used to determine the effective receiver sensitivity based upon the bandwidth of the received carrier, the noise figure of the receiver, and the manmade noise and interference in the environment.

  Downtilt calculator is used to calculate the antenna downtilt necessary to point the main beam of the antenna at the desired coverage area

  Path balance calculator and design signal strength objectives is used to determine the design signal strength requirements necessary to provide coverage on a street, in a vehicle, and in a building. It derives these signal requirements based upon the transmit power, receive sensitivity, antenna gain, and transmission system loss of the base station and CPE equipment.

  Watts to dBm calculator is used to convert power in watts to power in dBm and vice versa.

- FCC Documents

  Appendix A contains the FCC Part 15 rules as of March 2003. These rules cover the implementation requirements for systems (such as 802.11b and 802.11a) that operate in the unlicensed spectrum allocations. These rules cover the power and antenna limits associated with the unlicensed bands, and should be understood by anyone deploying equipment in these bands.

  FCC Part 15 NPRM is the September 15, 2003 Notice of Proposed Rule Making that proposes some changes to the Part 15 and Part 2 rules. These proposed changes are not yet (as of April 2004) written into law but if accepted would have a significant impact on the ability to utilize the unlicensed bands for community area service.

Code of Federal Regulations Part 27 is the FCC rule set pertaining to the bands under 5 GHz that can support nonline-of-sight fixed or mobile data services.

■ Measurement Tools

This folder contains copies of Kismet and Netstumbler. These programs can be used to allow a computer operating under Linux (Kismet) or Windows (Netstumbler) and equipped with a compatible 802.11 client device to be used as a signal strength and interference analysis tool. These tools are useful for site surveys and for troubleshooting of 802.11-based networks.

■ Operator Design and Financial Models

These large spreadsheets are examples of the modeling that can be done to analyze the size, complexity and cost of a large scale wireless network. Assigning values to key variables associated with equipment performance, cost, subscriber behavior, and the coverage area to be deployed allows the model to use area and demographic data associated with U.S. cities to determine the number of base station locations necessary to both cover the area and provide sufficient capacity to serve the subscribers based upon their average usage characteristics. The model can determine the capital and operating expense associated with the resulting network.

These models provide rough estimates of need only, and should be used for comparing equipment from different vendors, or for "what if" studies. They provide a way of rapidly evaluating changes to the system characteristics, rollout plan or subscriber usage changes. Final system design will require more granular terrain and demographic data which can be used by an RF engineer and a propagation analysis program to determine the RF performance of actual available site locations and the actual subscriber distribution within those coverage areas.

■ Radio Mobile Deluxe

This folder contains a copy of the Radio Mobile Deluxe propagation analysis software, and basic instructions for getting it loaded. This software uses the Longley-Rice propagation model to predict coverage using publicly available terrain data.

High-Speed Wireless Data:
System Types, Standards-Based and Proprietary Solutions

- Fixed Networks
- Nomadic Networks
- Mobile Networks
- Standards-Based Solutions and Proprietary Solutions
- Overview of the IEEE 802.11 Standard
- Overview of the IEEE 802.16 Standard
- 10–66 GHz Technical Standards
- 2–11 GHz Standards
- Overview of the IEEE 802.20 Standard
- Proprietary Solutions

High-Speed Wireless Data:
System Types, Standards-Based and Proprietary Solutions

Wireless data networks are often divided into several categories according to how the networks are viewed by the user. Such characteristics as fixed or mobile, point-to-point (PTP) or point-to-multipoint (PTM), licensed or unlicensed, and standards-based or proprietary are used to define the network. In reality, there are only two distinct types of networks: fixed or mobile. For purposes of definition fixed networks include networks that connect two or more stationary locations as well as systems like 802.11-based networks designed to support "nomadic" users. The nomadic user is nominally a fixed user constrained by the bounds of coverage available on the network. In a truly mobile system, the service will be ubiquitously available, and support use while the user is in motion. The first systems to offer true broadband mobile data are still years away at the time this book is being written. By adding EDGE, GPRS, 1XRTT, and 1XEVDO overlays to their voice networks, cellular and PCS carriers have taken the first tenuous steps in the direction of providing true mobile data, but the speeds at which current networks function cannot yet be called *broadband*. That designator can be used when the average connection speed per user exceeds 2 Mbps.

The systems discussed in this book will deliver true broadband connectivity. Available equipment can support speeds in excess of 500 Mbps. The equipment utilized in a network will be impacted by the type of network being implemented as well as the costs and service expectations of the network. While more complex networks require attention to more variables, RF design tools and knowledge requirements are fairly common for all networks regardless of their type.

Fixed Networks

The simplest network is the fixed point-to-point network. As the name implies, these are facilities that connect two or more fixed locations such as buildings. They are designed to extend data communications to locations physically separate from the rest of the network. A fixed network solution could be used to connect buildings together, to provide a network connection to a home, or to connect multiple network elements together.

These links may be familiar as the traditional microwave link. They use highly directional antennas in order to achieve range and control interference. Depending on the technology selected and the frequency of operation, these links can be designed to span distances as short as several hundred feet or as long as 20 or more miles, with capacities of under 1 Mbps to nearly 1 Gbps.

These systems are designed and engineered as individual radio paths, each path connecting two points together. A network of many of these individual paths could be designed to connect a multitude of disparate locations. For example, a fixed point-to-point network constructed out of a number of unique point-to-point links could be used to extend high-speed connectivity from a central point to a number of buildings in a campus or office park. It could also be used to extend the high speed connectivity of a fiber optic-based network to buildings surrounding the fiber route, thus avoiding the cost and complexity of digging up the streets to extend lateral connections from the fiber route into those other buildings.

Another variant of point-to-point networks are point-to-multipoint networks. In these networks a master or central station no longer uses individual antennas, each focused on a single station. Instead, it uses a wide aperture antenna that is capable of serving many stations in its field of view. In this way, a single system and antenna can share its capacity with a number of users. The benefit of such a system is that a single antenna can serve multiple locations, thus eliminating the need for many individual dish antennas to be located on the roof or tower that serves as the central location. The downside of such a network is threefold. Because the central station uses antennas that cover a wider area, they have a lower gain. This reduces the distance these networks can communicate as compared to a point-to-point network. Secondly, since many users share network capacity it may not be the optimal solution for supporting multiple very high bandwidth users. As with any system, the peak capacity requirements of the users must be considered as part of the overall network design. In the case of point-to-multipoint networks, the peak usage characteristics of multiple users must be considered. The third downside is related to interference management and frequency reuse. Since the central site transmits over a wide area, the ability to reuse the same frequencies in the network becomes more limited.

Point-to-point and point-to-multipoint networks can be accomplished using the licensed or unlicensed bands that exist in frequency ranges from under 1 GHz to over 90 GHz. They can use a multitude of proprietary technologies, or can be accomplished using equipment built to standards such as 802.11 or 802.16. Your selection of operating frequency and technology will be governed by factors such as range, capacity, spectrum availability, link quality, and cost.

Nomadic Networks

Another variation of point-to-multipoint networks is the network that directly supports a user's connection to the system. By this I mean instead of connecting buildings together, these nomadic networks connect individual computer users to the network. In the case of a laptop computer or PDA, these computing devices are somewhat mobile, and the network is designed to offer a low level of mobility to these users.

802.11b is a common standard for this type of network, although 802.11g and 802.11a also support this type of use. In order to be truly portable the RF device in the computer must be small, low powered, and the antennas used at the computer must be small and have an omnidirectional pattern. In addition, the user may be shielded from the base station by walls or other objects that attenuate the signal. This leads to a significant reduction in the area that can be effectively covered by one of these networks. Where point-to-point network range could be measured in miles, a nomadic implementation of the same technology has ranges measured in tens to hundreds of yards.

Nomadic networks are becoming quite commonplace. An 802.11b network offering Internet access in a coffee shop is one example of this type of network. Wireless office LANs, and WISP networks covering campuses or Multiple Dwelling Units (MDUs), like apartments, can also be considered nomadic networks.

These networks are the first step being taken to provide individuals with high-speed data access in many public and private venues.

These networks are not true mobile networks. While they can provide some mobility, they do not cover large areas and they do not support the high velocity mobility that would be needed to support a user in a vehicle. As with everything there are trade-offs. Localized low mobility solutions are fairly easy and inexpensive to implement. Better yet, there is unlicensed spectrum available to use for building this type of network, and a large installed base of customer equipment built to operate on the 802.11b Wi-Fi standard already exists. These factors have led to the rapid development of all sorts of nomadic networks, some as small as a home; others as large as a community.

Mobile Networks

The most complex network is one designed for true mobility. Like a voice-based cellular or PCS network, the high-speed mobile data network must provide ubiquitous coverage, and must support high velocity mobility. These requirements are not easily achieved or inexpensive. These systems will require many tens of megahertz of licensed spectrum, and will require technology that can deal with the hostile RF environment found in a truly mobile application. The 802.16e, 802.20 and CDMA2000 standards are several of the standards that may eventually bring true broadband mobile data solutions to large areas of the earth. Because of their cost, complexity, and need for interference managed dedicated spectrum, large telecom carriers, as opposed to the small businesses that offer nomadic network solutions, will be the most likely owner of these networks.

Standards-Based Solutions and Proprietary Solutions

The IEEE has a number of working groups responsible for developing open standards. These open standards are available for any manufacturer to use, hopefully ensuring competition and volume production. The IEEE has developed the 802.11x and 802.16 standards, and as of July 2003 has a working group developing the 802.20 standard.

Each of these standards is designed with a certain utility and limitations in mind. For example 802.11b was designed as a short-range wireless Ethernet replacement. While it can be used for other applications (such as community networks) it is not optimized for this type of service, and will never perform as well as a technology that was designed from the ground up to address the unique issues found in a community network.

Certain manufacturers develop equipment that is not designed to any current IEEE or other standard. These solutions sometimes become popular enough that they become a de facto standard. More often, these proprietary standards become niche market solutions, which are only available from a single source. These proprietary solutions may be technically best suited for certain applications, but often are significantly more expensive than standards-based options. In the end, it's up to you to determine whether the improved performance is worth the additional cost and single-vendor supply risks.

There are many proprietary solutions available; unfortunately in a competitive market manufacturers are not willing to release much detail about their equipment operation and performance without the recipient signing a Nondisclosure Agreement (NDA) which limits the amount of information that can be shared or published. Because of this limitation I will not be spending much time discussing proprietary solutions in detail.

Overview of the IEEE 802.11 Standard

Like many standards, 802.11 has gone through many iterations and expansions over the years. Initially encompassing a 1 Mbps throughput on a 900 MHz channel, it now supports up to 54 Mbps in the 2400 MHz and 5600 MHz bands.

802.11x, also sometimes known as Wi-Fi, is an IEEE certified wireless networking standard that currently includes the IEEE 802.11a, 802.11b and 802.11g specifications. In the U.S., the RF emission of these devices is governed by FCC Part 15 rules. These rules govern the power output, equipment and antenna configurations useable in the unlicensed bands. A copy of the FCC Part 15 rules is included on the CD-ROM that accompanies this book. The 802.11b spec allows for the wireless transmission of approximately 11 Mbps of raw data at indoor distances from several dozen to several hundred feet and outdoor distances of several to tens of miles as an unlicensed use of the 2.4 GHz band. The 802.11a spec uses the unlicensed 5 GHz band, and can handle 54 Mbps over shorter distances.

The 802.11g standard applies the 802.11a modulation standards (and therefore supports 54 Mbps just like 802.11a) to the 2.4 GHz band, and offers "backward compatibility" for 802.11b devices. The achievable coverage distances for these standards depend on impediments and obstacles to line of sight.

The 802.11b specification started to appear in consumer form in mid-1999, with Apple Computer's introduction of its AirPort components, manufactured in conjunction with Lucent's WaveLAN division. (The division changed its named to Orinoco and was spun off to the newly formed Agere Corporation with a variety of other Lucent assets in early 2001; these assets were resold to Proxim Corporation in June 2002, although Agere continues to make chips.)

802.11x is an extension of wired Ethernet, bringing Ethernet-like principles to wireless communication. As such, 802.11 is agnostic about the kinds of data that pass over it. It's primarily used for TCP/IP, but can also handle other forms of networking traffic, such as AppleTalk or NetBEUI.

Computers and other devices using Windows or Mac OS operating systems, and many flavors of Unix and Linux, may communicate over Wi-Fi, using equipment from a variety of vendors. The client hardware is typically a PC card or a PCI card, although USB and other forms of Wi-Fi radios are also available. Adapters for PDAs, such as Palm OS and PocketPC based devices, are available in various forms, and smaller ones that fit into internal Secure Digital and Compact Flash card slots started appearing in late 2002.

Each radio may act, depending on software, as a hub or as part of an ad hoc computer-to-computer transmission network; however it's much more common that a Wireless Local Area Network (WLAN) installation uses one or more Access Points (AP), which are dedicated stand-alone hardware with typically more powerful chipsets and higher gain antennas. Home and small-office APs often include routing, a DHCP server, NAT, and other features required to implement a simple network; enterprise access points include access control features as well as secure authentication support.

The 802.11b standard as implemented in the 2.4 GHz band is backwards compatible with early 2.4 GHz 802.11 equipment. 802.11b can support speeds of 1, 2, 5.5 and 11 Mbps on the same hardware. Multiple 802.11b access points can operate in the same overlapping area over different channels, which are subdivisions of the 2.4 GHz band available.

Internationally, there are 14 standard channels, which are spaced at 5 MHz intervals, from 2.4000 to 2.487 GHz. Only channels 1 through 11 are legal in the U.S.A. The 802.11 channel is 22 MHz wide, so it occupies multiple 5 MHz channels (see Figure 1-1). Only channels 1, 6, and 11 can be assigned to an 802.11 network with no overlap among them. If closer spaced channels are assigned, there will be inter-carrier interference generated. Such overlapping systems can still work, but the interchannel interference will effectively raise the noise floor in the channel, which will have a negative impact on the throughput and range of the systems.

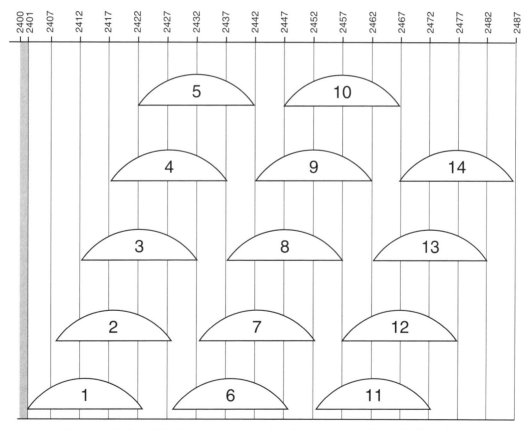

Figure 1-1: 2.4 GHz frequency chart showing 802.11b channel allocations

802.11b uses several types of modulation. Barker Code Direct Sequence Spread Spectrum with BPSK or QPSK modulation is used to transmit at 1 and 2 Mbps respectively, while Complimentary Code Keying is used to support speeds of 5.5 and 11 Mbps. Multiple users are supported by the use of Carrier Sense Multiple Access/Collision Avoidance (CSMA/CA).

A new higher speed standard called *802.11g* features complete backwards compatibility with 802.11b, but it offers three additional encoding options (one mandatory, two optional) that boost its speed to 54 Mbps, although two 22 Mbps "flavors" are part of the specification as well. The higher speed connections use the same modulation as 802.11a: Orthogonal Frequency Division Multiplexing (OFDM). Future speed improvements achieved through the use of more efficient modulations are expected in 802.11 products operating in both the 2.4 and 5 GHz bands.

802.11a specifies the use of OFDM modulation only, and supports data rates of 6, 9, 12, 18, 24, 36, 48, or 54 Mbps of which 6, 12, and 24 Mbps are mandatory for all products. OFDM operates extremely efficiently, thus leading to the higher data rates. OFDM divides the data signal across 48 separate sub-carriers to provide transmissions. Each of the sub-carriers uses phase shift keying (PSK) or Quadrature Amplitude Modulation (QAM) to modulate the digital signal depending on the selected data rate of transmission. In addition, four pilot sub-carriers provide a reference to minimize frequency and phase shifts of the signal during transmission.

Multiple users are supported by the use of CSMA/CA, so the same limitations inherent in this access methodology for 802.11b will be present in 802.11a as well.

The operating frequencies of 802.11a fall into the U-NII bands: 5.15–5.25 GHz, 5.25–5.35 GHz, and 5.725–5.825 GHz. As shown in Figure 1-2, within this spectrum there are twelve 20 MHz channels (eight allowable only for indoor use and four useable for indoor or outdoor use) that do not overlap, thus allowing denser installations. Additionally, each band has different output power limits that are detailed in the FCC rules Part 15.407. 802.11a's range is less due to both its frequency of operation and more complex modulation, but in a closed environment like an office or a home it can often transmit higher speeds at similar distances as compared to 802.11b.

As 802.11 systems have proliferated, a number of issues surrounding its limitations have been raised. The security provisions of 802.11 are notoriously weak,

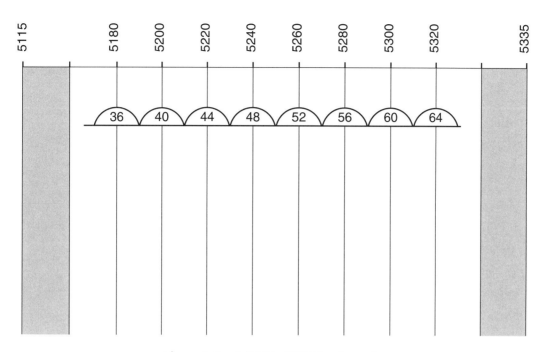

Figure 1-2a: 5.2 GHz 802.11a channels

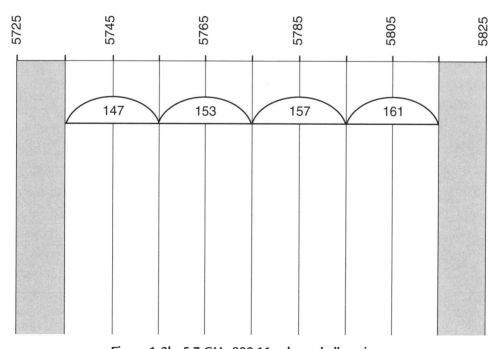

Figure 1-2b: 5.7 GHz 802.11a channel allocations

and it does not inherently support Quality of Service packet prioritization. New working groups at the IEEE are addressing these limitations. The 802.11e, h, and i standards will improve the capabilities of 802.11, and make it more robust and more useful in a number of situations.

Even though standards exist, they do not guarantee that equipment from different manufacturers will interoperate. To assure interoperability, and thereby assist adoption by the consumer, an industry group known as the Wi-Fi Alliance (formerly known as the Wireless Ethernet Compatibility Alliance) certifies its members equipment as conforming to the 802.11a and b standards, and allows compliant hardware to be stamped Wi-Fi compatible. The Wi-Fi seal of approval is an attempt at a guarantee of compatibility between hundreds of vendors and thousands of devices. (The IEEE does not have such a mechanism, as it only promulgates standards.) In early October 2002, the group modified the Wi-Fi mark to indicate both a and b standards by noting 2.4- or 5-GHz band compatibility.

802.11b was the first standard deployed for public short-range networks, such as those found at airports, hotels, conference centers, and coffee shops and restaurants. Several companies currently offer paid hourly, session-based, or unlimited monthly access via their deployed networks around the U.S. and internationally.

802.11a and b are a great way to extend a data or Internet connection to a site that does not have one through point-to-point operation, or to build a point-to-multipoint system which could provide a high speed data connection shared by a number of fixed and nomadic users.

As delivered, 802.11 products conform to FCC Part 15 rules, which limit both the device RF power and EIRP achieved by use of a gain antenna. The most stringent restrictions are placed on omnidirectional operations, since those operations result in the highest overall interference contribution to the surrounding area. In the case of omni operation, the EIRP is limited to 1 watt. If a directional antenna is used, the allowable EIRP jumps to 4 watts. If a fixed point-to-point link is implemented, even higher EIRP is available, in fact with 30 dBi gain antennas, over 100 watts EIRP can be achieved. EIRP of this magnitude will support a point-to-point link over 15 miles long, given the right conditions in the path.

802.11 does have a significant impediment when used in a WISP or MAN type deployment: like the Ethernet standard upon which it is based, it uses Carrier

Sense Multiple Access as its access protocol. In the case of 802.11, the full specification is Carrier Sense Multiple Access/Collision Avoidance (CSMA/CA), which anticipates that all stations are able to hear each other, and thus to have the ability to listen for activity on the channel prior to transmitting. If another carrier is heard, the station knows the channel is in use and "backs off" for a random time. At the end of the back off interval, the station listens again, and transmits if the channel is clear.

In a widespread outdoor system the end users can hear the base station or AP, but they may not be able to hear each other, so CSMA/CA does not work. As the system gets loaded it is possible for multiple users all to transmit at once, leading to interference and packet loss. Another effect is the "near far" issue in which users closer to the base station get better access to the bandwidth than stations further away. This is due to the higher signal strength of the close-in user swamping the weaker signal of the remote user. Enabling Request to Send/Clear to Send (RTS/CTS) can diminish these effects. With RTS/CTS enabled, the stations ask permission before transmitting, and wait to get an all clear before they actually start their transmission. Still, there is no way for a station to know whether a CTS was in response to his RTS or another sent by another station, so collisions can still happen. Additionally, RTS/CTS adds additional overhead, thus reducing the bandwidth available to carry real traffic.

Several manufacturers have addressed these issues by adding some semblance of central control to the systems. By queuing the users and specifying who gets to transmit next, there is an improvement in usability, quality and throughput for all users. Of course, such equipment uses proprietary standards, and some even require the radio hardware to be their proprietary solution.

802.11 was never designed for WISP or Metropolitan Area Network (MAN) deployment. It has become a de facto standard for such use because it is inexpensive, has freely available spectrum, and works well enough on lightly loaded systems. The IEEE recognized the limitations of 802.11 and the need for a MAN solution. The 802.16 standard was promulgated in order to provide a solution for the WISP and metropolitan area network provider. Because 802.16 was designed to cover large areas, its MAC layer does not use CSMA/CA, so it does not exhibit 802.11's CSMA/CA limitations.

Overview of the IEEE 802.16 Standard

The IEEE 802.16 Air Interface Standard is a state-of-the-art specification for fixed broadband wireless access systems employing a Point-to-Multipoint (PMP) architecture. The initial version was developed with the goal of meeting the requirements of a large number of deployment scenarios for Broadband Wireless Access (BWA) systems operating between 10 and 66 GHz. As a result, only a subset of the functionality is needed for typical deployments directed at specific markets.

An amendment to add support for systems operating between 2 and 11 GHz was added to the initial specification. Since the IEEE process stops short of providing conformance statements and test specifications, in order to ensure interoperability between vendors the WiMAX forum was created. In much the same manner as the Wi-Fi forum assured equipment interoperability to the 802.11 standard, the WiMAX forum will provide the testing and certification necessary to assure vendor equipment interoperability for 802.16 hardware.

Task Group 1 of the IEEE 802.16 committee developed a point-to-multipoint broadband wireless access standard for systems in the frequency range 10–66 GHz. The standard covers both the Media Access Control (MAC) and the PHYsical (PHY) layers. Task groups a and b jointly produced an amendment to extend the specification to cover both the licensed and unlicensed bands in the 2–11 GHz range.

A number of PHY considerations were taken into account for the target environment. For example at frequencies above 6 GHz, line of sight (LOS) paths between stations are a must. By taking the need for line of sight paths as a design requirement the PHY can be designed for minimal effects related to multipath. This allows the PHY to accomodate wide channels, typically greater than 10 MHz in bandwidth, thus giving IEEE 802.16 the ability to provide very high capacity links on both the uplink and the downlink.

At the lower frequencies, line of sight is not required for link operation, although the lack of Line of Sight (LOS) forces other design trade-offs. Adaptive burst profiles (changing both modulation and Forward-Error Correction (FEC)) are used to further increase the typical capacity of 802.16 systems with respect to older technology. The MAC was designed to accommodate different PHYs for the different environments. The single-carrier PHYs are designed to accommodate either Time Division Duplexing (TDD) or Frequency Division Duplexing (FDD) deployments, allowing for both full and half-duplex terminals in the FDD case.

The MAC was designed specifically for the PMP wireless access environment. It is designed to carry any higher layer or transport protocol such as Asynchronous Transfer Mode (ATM), Ethernet or Internet Protocol (IP) seamlessly, and is designed to accommodate future protocols that have not yet been developed. The MAC is designed for the very high bit rates (up to 268 Mbps each way) of the truly broadband physical layer, while delivering ATM compatible Quality of Service (QoS) to ATM as well as non-ATM (MPLS, VoIP, and so forth) services.

The frame structure allows terminals to be assigned uplink and downlink burst profiles dynamically according to their link conditions. This allows a trade-off between capacity and robustness in real-time, and provides an approximate twofold increase in capacity on average when compared to nonadaptive systems, while maintaining appropriate link availability.

The 802.16 MAC uses a variable length Protocol Data Unit (PDU) along with a number of other concepts that greatly increase the efficiency of the standard. Multiple MAC PDUs may be concatenated into a single burst to save PHY overhead. Additionally, multiple Service Data Units (SDU) for the same service may be concatenated into a single MAC PDU, thus saving on MAC header overhead. Variable fragmentation thresholds allow very large SDUs to be sent piece meal to guarantee the QoS of competing services. Additionally, payload header suppression can be used to reduce the overhead caused by the redundant portions of SDU headers.

The MAC uses a self-correcting bandwidth request/grant algorithm known as Demand Assigned Multiple Access/Time Division Multiple Access (DAMA/TDMA) that eliminates the shortcomings of the CSMA/CA technique. DAMA adapts as needed to respond to demand changes among multiple stations. With DAMA, the assignment of timeslots to channels varies dynamically based upon need. For transmission from a base station to subscribers, the standard specifies two modes of operation, one targeted to support a continuous transmission stream (mode A), such as audio or video, and one targeted to support a burst transmission stream (mode B), such as IP-based traffic. User terminals have a variety of options available to them for requesting bandwidth depending upon the QoS and traffic parameters of their services. Users can be polled individually or in groups. They can signal the need to be polled, and they can piggyback requests for bandwidth.

10–66 GHz Technical Standards

In the same manner as the Wi-Fi consortium managed compatability for 802.11 devices, the WiMAX forum is working with 802.16 products and standards to assure broad compatability. Since the 10–66 GHz standard was the first to be released, WiMAX initially created a 10–66 GHz technical working group. The technical working group created equipment operating profiles and test specifications, but an authorized, independent laboratory does actual testing. For each system profile, functions are separated between mandatory and optional feature classes. There can be differences from one equipment manufacturer to another in implementing optional features, but mandatory features will be the same in every vendor's product.

WiMAX is currently defining two MAC system profiles, one for basic ATM and the other for IP-based systems. Two primary PHY system profiles are also being defined: a 25 MHz-wide channel (typically for U.S. deployments) for use in the 10–66 GHz range and a 28 MHz wide channel (typically for European deployments) also for use in the 10–66 GHz range. The PHY profiles are identical except for their channel width and their symbol rate, which is proportional to their channel width. Each primary PHY profile has two duplexing scheme sub-profiles one for Frequency Division Duplex (FDD) and another for Time Division Duplex (TDD). Additionally, because these systems were designed for operating over LOS paths, traditional multistate QAM modulation is used.

2–11 GHz Standards

In early 2003, the IEEE 802.16 standard was expanded with the adoption of the 802.16a amendment, focused on broadband wireless access in the frequencies from 2 to 11 GHz. Given the charter of the WiMAX forum, to promote certification and interoperability for microwave access around the globe, WiMAX has expanded its scope to include the 802.16a standard.

The 802.16a standard is designed to operate over both LOS and NLOS paths. Because of the multipath effects present in NLOS paths, QAM as used in the

10–66 GHz 802.16 variant, was not a suitable modulation. 802.16a instead uses OFDM as its modulation technique.

The WiMAX 2–11 GHz working group is currently defining MAC and PHY System profiles for IEEE 802.16a and HiperMAN standards. The MAC profiles that are being developed include IP-based versions for deployment in both licensed and unlicensed spectrum.

While the IEEE 802.16a amendment has several physical layer profiles, the WiMAX forum is focusing on the 256-point FFT OFDM PHY mode as its initial and primary interoperability mode. Various channel sizes that cover typical spectrum allocations in both licensed and license exempt bands around the globe have been chosen. All selected channel sizes support the 256-point FFT OFDM PHY mode of operation.

In February 2003, the IEEE instituted another working group, the 802.16e working group. The 802.16e extension adds vehicular speed mobility in the 2 to 6 GHz licensed bands. At the time this book is being written, this extension is still in committee. It is anticipated that the standard will be released in mid 2004.

Overview of the IEEE 802.20 Standard

The 802.20 standard focuses on true high velocity mobile broadband systems. The 802.20 interface seeks to boost real-time data transmission rates in wireless metropolitan area networks to speeds that rival DSL and cable modem connections (1 Mbps or more). This will be accomplished with base stations covering radii of up to 15 kilometers or more, and it plans to deliver those rates to mobile users even when they are traveling at speeds up to 250 kilometers per hour (155 miles per hour). This would make 802.20 an option for deployment in high-speed trains. The standard is focused on operation in licensed bands below 3.5 GHz.

The 802.20 Working Group was actually established before the IEEE gave the go-ahead to 802.16e. The IEEE originally intended to have the 802.20 standard in place by the end of 2004, but the group has been mired in conflict and has made little progress to date.

802.20 may become a direct competitor to third-generation (3G) wireless cellular technologies such as CDMA2000 and GMRS. Instead of using TDMA or CDMA technology, 802.20 is expected to use OFDM as its modulation technique.

Proprietary Solutions

In addition to the standards-based solutions, there are numerous vendor proprietary systems available. Proprietary solutions are normally designed to best suit a particular deployment scenario, and may operate in licensed bands, unlicensed bands, or in some cases both.

Just like standard solutions, proprietary solutions continue to evolve in order to keep a competitive edge and to better meet the needs of a growing business opportunity. Because of the financial dynamics associated with companies providing proprietary solutions as well as the changing requirements of the marketplace, there is no guarantee that the equipment or manufacturers discussed next will still be available by the time you read this. Table 1-1 lists some of the proprietary manufacturers and the publicly available product specifications that were available in late 2003.

Because proprietary solutions are just that: proprietary, it is often extremely difficult to obtain specific information about the operation of the hardware without signing a nondisclosure agreement with the vendor. Of course this is only possible if the vendor will commit to such an agreement with you. Lacking these particulars about the equipment can make it difficult to compare operating characteristics of the equipment, and analyze how a particular solution might fulfill your particular requirements. Happily, all is not lost. Generally, information about capacity and throughput is generally publicly available. The missing information usually relates to the actual RF operating characteristics of the hardware.

In order for any radio transmitting equipment to be sold in the U.S., it is required to go through an FCC certification process. This certification is accomplished by an independent testing lab, which conducts tests and measurements on the equipment to assure that it meets the FCC's technical requirements for the band in which it operates. The result of passing this certification process is that the FCC grants an authorization number, which is used by the manufacturer to show that the equipment is legally operating within the FCC rules, and that it can legally be sold for operation in the U.S.

The FCC publishes the results of these tests as public record. They can be found at the FCC Equipment Authorization System Generic Search web page. As of

February, 2004 this page is located at: https://gullfoss2.fcc.gov/prod/oet/cf/eas/reports/GenericSearch.cfm. Knowing as little as the manufacturers name and the band of operation will allow you to use this search engine to identify certified equipment. Once you've identified the particular equipment you're interested in, read the test results and other documentation on file. While some information may be held in confidence, important information like spectrum analyzer plots of the output waveform and the output power will be part of the public record. Knowing power output is key to analyzing the coverage you can expect from a particular solution. Receiver sensitivity is the other factor you will need to learn. In many cases this is not a published part of the FCC certification tests, so you will need to derive it from the information available. The waveform is a useful tool for this investigative work. It will offer clues about the modulation in use and the spectral occupancy of the signal. Couple that with information about the capacity or throughput of the solution, and you can begin to understand the operating characteristics of the receiver. An RF hardware design expert could use this information to derive the receiver sensitivity based on the characteristics. Alternately estimation could be made by assuming that the sensitivity will be similar to that of known equipment having similar modulation characteristics.

Now that you have a brief overview of the types of solutions available for use in deploying a wireless data network, it's time to move on to gaining an understanding about how radio works and what issues must be considered in designing a radio-based network.

Table 1-1

Company	Product	Operating Band	Total Speed	Maximum number of Sectors	Maximum number of users per Sector	LOS/NLOS	Modulation	Encryption Levels	Output Power	Receiver Sensitivity	Point to Point	Point to Multipoint	Duplex
AIRAYA	AJ108 Wireless Bridge	5.25–5.35 GHz	108 Mbps	1	1	LOS	OFDM	152-bit	EIRP 29.6 dB		Yes	No	Half
AIRAYA	AJ108 Wireless Bridge	5.25–5.35 GHz	108 Mbps	1	1	LOS	OFDM	152-bit	EIRP 29.6 dB		Yes	No	Half
Alvarion	BreezeACCESS	2.4, MMDS, 3.5, 5.15–5.35, 5.4, 5.7 UNII, 5.7 ISM	3 Mbps for GFSK, 12 Mbps for 3.5 GHz	depends on band, up to 12 in 2.4, up to 36 in 3.5 (with multibeam)	1000	LOS and NLOS options	FHSS, DSSS, & OFDM	128-bit, triple DES option	26 dBm, with APC		Yes	Yes	TDD
Aperto	PacketWave	2.5 GHz, 3.5 GHz and 5.8 GHz	20 Mbps per sector	6	1000	LOS/NLOS		Other	Varies depending on band		No	Yes	Half
Aperto	PacketWave Point-to-Point Bridges	5.8 GHz	20 Mbps per sector	1	1	LOS/NLOS		Other	Varies depending on band		Yes	No	Half
Axxcelera	AB-Access	5.7 UNII band	25 Mbps PMP, 12.5 Mbps PTP	12	256	LOS	TDMA/TDD	Yes, proprietary	32 dBm	–86.1 dB for 10–4 BER	Yes	Yes	full or half
BeamReach	BeamPlex	2.3 GHz		Not Applicable	16,000 per cell	NLOS	Adaptive Multi-Beam OFDM	Yes				Yes	
BeamReach	BeamPlex	2.3 GHz		Not Applicable	16,000 per cell	NLOS	Adaptive Multi-Beam OFDM	Yes				Yes	
Ceragon	FibeAir	6,7,8,11,13,15,18, 23,26,28,31, 32,38 GHz	622 Mbps	N/A	N/A	LOS		DES	MAX 24 dBm	–73 dBm	Yes	No	Full
Cirronet	WaveBolt	2.4 GHz and 5.8 GHz (5.8 available in Q1 '03)	1 Mbps	5	240	LOS/NLOS	FHSS	Proprietary	+18 dBm	–88 dBm	Yes	Yes	Full
Dragonwave	AirPair	18,23,28 GHz	50–100 Mbps	N/A	N/A	LOS	Single Carrier QAM	REL 3 will include DES encryption	+13/+17 dBm	–77/–80.5 dBm	Yes	N/A	Full
Innowave	MGW	0.8; 1.5; 1.9; 2.4; 3.4–3.8 GHz	850 Kbps	6	1000	NLOS	FHCDMA/ TDMA/TDD	Other	27 dBm	–90 dBm	No	Yes	Full
Innowave	eMGW	1.5; 1.9; 2.4; 3.4–3.8; 5.7 GHz	1.5 Mbps	6	2000	NLOS	FHCDMA/ TDMA/TDD	Other	27 dBm	–90 dBm	No	Yes	Full
Innowave	WaveGain	3.5 GHz	15 Mbps	4	128	NLOS	WCDMA/FDD	Other	37 dBm	–110 dBm	No	Yes	Full
MeshNetwork	MEA IAP6300	2.4 GHz	6 Mbps	N/A	250	NLOS	Multi-Hop DSSS	*VPN, IPSEC	22+ dBm		Yes	Yes	Half

(continued)

Table 1-1

Company	Product	Operating Band	Total Speed	Maximum number of Sectors	Maximum number of users per Sector	LOS/NLOS	Modulation	Encryption Levels	Output Power	Receiver Sensitivity	Point to Point	Point to Multipoint	Duplex
Motorola	Canopy	5.2 GHz	10 Mbps	6	200	LOS		single DES	30 dB	–83 dB	Yes	Yes	Half
Motorola	Canopy	5.7 GHz	10 Mbps	6	200	LOS		single DES	30 dB	–83 dB	Yes	Yes	Half
Navini	Ripwave	2.4 GHz; 2.5/2.6 GHz; 2.3 GHz	48 Mbps (3.3 sectored cell), 72 Mbps in 2003	3	1000	NLOS, Zero-install plug-and-play	Phased-array smart antennas, Multi-Carrier Synchronous Beamforming (MCSB), TDD	Patented CDMA encoding + spatial isolation and nulling with beamforming provides for a high level of security and can be overlayed.	Depends on the frequency of operation		Yes	Yes	Full
Nokia	Nokia RoofTop	2.4 GHz	12 Mbps aggregate	6	40	NLOS	FHSS	none	12 dBm to 27 dBm	–82 dBm	No	No	Half duplex
P-Com	AirPro Gold.Net	2.4 & 5.8 GHz	11 Mbps	4	127	Both	FHSS	MAC security	+28 dBm – 2.4 GHz, +27 5.8 GHz	85 dBm @10^6 BER	Yes	Yes	Full
Proxim	Tsunami Multipoint 20 MB Base Station Unit	5.8 GHz	20 Mbps	6 typical max per hub site	1,023	Near Line of Sight		Proprietary	36 dBm		No	Yes	Half
Proxim	Tsunami Multipoint 20 MB Subscriber Unit	5.8 GHz	20 Mbps	6 typical max per hub site	N/A	Near Line of Sight		Proprietary	35 dBm		No	Yes	Half
Proxim	Tsunami Multipoint 60 MB Base Station Unit	5.8 GHz	60 Mbps	6 typical max per hub site	1,023	Near Line of Sight		Proprietary	36 dBm		No	Yes	Half
Proxim	Tsunami Multipoint 60MB Subscriber Unit	5.8 GHz	60 Mbps	6 typical max per hub site	N/A	Near Line of Sight		Proprietary	35 dBm		No	Yes	Half
Proxim	QuickBridge 20	5.8 GHz	18 Mbps aggregate capacity	N/A		LOS		16 Char Security ID	+36 dBm EIRP	–89 dBm	Yes	No	Half duplex
Proxim	QuickBridge 60	5.8 GHz	54, 36, 18 Mbps aggregate capacity	N/A		LOS		16 Char Security ID	+36 dBm EIRP	–77 dBm	Yes	No	Half duplex
Proxim	QuickBridge 20 +T1	5.8 GHz	12 Mbps aggregate capacity	N/A		LOS		16 Char Security ID	+36 dBm EIRP	–89 dBm	Yes	No	Half duplex

(continued)

Table 1-1

Company	Product	Operating Band	Total Speed	Maximum number of Sectors	Maximum number of users per Sector	LOS/NLOS	Modulation	Encryption Levels	Output Power	Receiver Sensitivity	Point to Point	Point to Multipoint	Duplex
Proxim	QuickBridge 20 +E1	5.8 GHz	12 Mbps aggregate capacity	N/A		LOS		16 Char Security ID	+36dBm EIRP	-89dBm	Yes	No	Half duplex
Proxim	Tsunami 10 2.4 GHz Wireless Ethernet Bridge	2.4 GHz	10 Mbps full duplex with wayside T1 channel	N/A		LOS	DSSS	8 bit Security Address	+27dBm	-86dBm	Yes	No	
Proxim	Tsunami 10 5.8 GHz Wireless Ethernet Bridge	5.8 GHz	10 Mbps full duplex with wayside T1 channel	N/A		LOS	DSSS	8 bit Security Address	+20dBm	-84dBm	Yes	No	
Proxim	Tsunami 45 5.8 GHz Wireless Fast Ethernet Bridge	5.8 GHz	45 Mbps full duplex with wayside T1 channel	N/A		LOS		12 char Security Code	+17dBm	-79dBm	Yes	No	
Proxim	Tsunami 45 5.3 GHz Wireless Fast Ethernet Bridge	5.3 GHz	45 Mbps full duplex with wayside T1 channel	N/A		LOS		12 char Security Code	+13dBm	-79dBm	Yes	No	
Proxim	Tsunami 100 5.3/5.8GHz Wireless Fast Ethernet Bridge	5.3 and 5.8 GHz	100 Mbps full duplex with wayside T1 channel	N/A		LOS		12 char Security Code	+10 and +17dBm	-77dBm	Yes	No	
Proxim	Tsunami 100 5.8 GHz Wireless Fast Ethernet Bridge	5.8 GHz	100 Mbps full duplex with wayside T1 channel	N/A		LOS		12 char Security Code	+16dBm	-71dBm	Yes	No	
RadioLAN	Campus BridgeLINK-II	5.775 GHz 5.3 GHz 5.2 GHz	10 Mbps	6	128	LOS	D-PPM	*WEP 128-bit *WEP 256-bit *Other	+17dBM		Yes	Yes	
RadioLAN	Campus BridgeLINK-Lite	5.775 GHz	10 Mbps	1	128	LOS	D-PPM	*WEP 128-bit *WEP 256-bit *Other	+17dBM		Yes	Yes	
RadioLAN	Campus BridgeLINK-II (RMG-377-EA1)	5.775 GHz 5.3 GHz 5.2 GHz	10 Mbps	1	128	LOS	D-PPM	*WEP 128-bit *WEP 256-bit *Other	+17dBM		Yes	Yes	
RadioLAN	Campus BridgeLINK-II (RMG-377-25P)	5.775 GHz 5.3 GHz 5.2 GHz	10 Mbps	1	128	LOS	D-PPM	*WEP 128-bit *WEP 256-bit *Other	+17dBM		Yes	Yes	
RadioLAN	Campus BridgeLINK-II (RMG-377-RW1)	5.775 GHz 5.3 GHz 5.2 GHz	10 Mbps	1	128	LOS	D-PPM	*WEP 128-bit *WEP 256-bit *Other	+17dBM		Yes	Yes	

(continued)

Table 1-1

Company	Product	Operating Band	Total Speed	Maximum number of Sectors	Maximum number of users per Sector	LOS/NLOS	Modulation	Encryption Levels	Output Power	Receiver Sensitivity	Point to Point	Point to Multipoint	Duplex
RadioLAN	Campus BridgeLINK-II (RMG-377-RW2)	5.775 GHz 5.3 GHz 5.2 GHz	10 Mbps	1	128	LOS	D-PPM	*WEP 128-bit *WEP 256-bit *Other	+17 dBM		Yes	Yes	
RadioLAN	Campus BridgeLINK-II (RMG-377-RW3)	5.775 GHz 5.3 GHz 5.2 GHz	10 Mbps	1	128	LOS	D-PPM	*WEP 128-bit *WEP 256-bit *Other	+17 dBM			Yes	
RadioLAN	Campus BridgeLINK-II (RMG-377-S90)	5.775 GHz 5.3 GHz 5.2 GHz	10 Mbps	4	128	LOS	D-PPM	*WEP 128-bit *WEP 256-bit *Other	+17 dBM		Yes	Yes	
Redline Communications	AN 100	3.4000- 3.800 GHz	Up to 70 Mbps			NLOS	OFDM		+23 dBM'	−88 dBm@ 1E-09 BER in a 7 MHz channel	Yes	Yes	TDD
Remec	ExcelAir® 70	3.5 GHz	Up to 300 Mbps	6	200-400	LOS	SCQAM	None	+40 dbi EIRP		No	Yes	Full
Solectek	SkyWay-NET	2.4 GHz	11 Mbps	6	64	LOS	DSSS	Proprietary	26 dBm	−83 dBm	Yes	Yes	Half
Solectek	SkyWay-LINK	2.4 GHz	11 Mbps			LOS	DSSS	Proprietary	26 dBm	−83 dBm	Yes	No	Half
Solectek	SkyMate CPE	2.4 GHz	11 Mbps	6	64	LOS	DSSS	Proprietary	23 dBm	−80 dBm	Yes	Yes	Half
Solectek	AIRLAN Bridge Kit	2.4 GHz	11 Mbps			LOS	DSSS	Proprietary	23 dBm	−80 dBm	Yes	No	Half
Solectek	AIRLAN Bridge 5	5.8 GHz	11 Mbps			LOS	DSSS	Proprietary	23 dBm		Yes	No	Half
Wi-LAN	AWE 120–24 Wireless Ethernet Bridge	2.4 GHz	12 Mbps raw, up to 9 Mbps effective	3	1000	LOS	DSSS	Proprietary up to 20 Byte	20 dBm	−81 dBm	Yes	Yes	Half Duplex
Wi-LAN	AWE 45–24 Wireless Ethernet Bridges	2.4 GHz	4.5 Mbps raw, up to 3.4 Mbps effective	3	250	LOS	DSSS	Proprietary up to 20 Byte	20 dBm	−83 dBm	Yes	Yes	Half Duplex
Wi-LAN	VIP 110-24	2.4 GHz	11 Mbps raw, up to 8 Mbps effective	4	500	NLOS	DSSS	Proprietary up to 20 Byte	0 to +23 dBm	−82 dBm	Yes	Yes	Half Duplex
Wi-LAN	Ultima3 RD (Rapid Deployment)	5.8 GHz	12 Mbps raw, up to 10 Mbps effective	N/A	N/A	LOS	DSSS	Proprietary up to 20 Byte	−10 dBm to +21 dBm	−80 dBm	Yes	No	Half Duplex

(continued)

Table 1-1

Company	Product	Operating Band	Total Speed	Maximum number of Sectors	Maximum number of users per Sector	LOS/NLOS	Modulation	Encryption Levels	Output Power	Receiver Sensitivity	Point to Point	Point to Multipoint	Duplex
Wi-LAN	Ultima3 ER (Extended Range)	5.8 GHz	12 Mbps raw, up to 10 Mbps effective	N/A	N/A	LOS	DSSS	Proprietary up to 20 Byte	-10 dBm to +21 dBm	-80 dBm	Yes	No	Half Duplex
Wi-LAN	Ultima3 AP (Access Point)	5.8 GHz	12 Mbps raw, up to 10 Mbps effective	4	1000	LOS	DSSS	Proprietary up to 20 Byte	-10 dBm to +21 dBm	-80 dBm	No	Yes	Half Duplex
Wi-LAN	Ultima3 CPE (Customer Premises Equipment)	5.8 GHz	12 Mbps raw, up to 10 Mbps effective	N/A	N/A	LOS	DSSS	Proprietary up to 20 Byte	-10 dBm to +21 dBm	-80 dBm	No	Yes	Half Duplex
Wi-LAN	LIBRA Series (Access Point)	3.5 GHz	16 Mbps raw, up to 12 Mbps effective (7 MHz)	6	2047	NLOS	W-OFDM		avg/peak +22/+32 dBm	-82 dBm/ -80 dBm (3.5/7 MHz)	Yes	Yes	Full Duplex
Wi-LAN	LIBRA Series (CPE)	3.5 GHz	16 Mbps raw, up to 12 Mbps effective (7 MHz)	N/A	N/A	NLOS	W-OFDM		+17/+27 dBm	-79dBm	No	Yes	Half Duplex

CHAPTER 2

Basic Radio and RF Concepts

- RF Energy
- RF Generation and Transmission
- Oscillator
- Power Amplifiers
- Antennas and Feedlines
- RF Reception
- Modulation of RF Signals
- Amplitude Modulation
- Frequency Modulation
- Complex Modulation
- Duplexing
- Frequency Division Duplexing
- Time Division Duplexing
- Multiple Access Techniques
- Spread-Spectrum Modulation
- OFDM
- Ultra Wideband

Basic Radio and RF Concepts

This chapter is intended to provide an overview of the subsystems that make up a radio system. By design, it will be simplistic in its approach. The intent is only to familiarize you with the basic operation of the building blocks of a radio-based system. Each of these building blocks is in reality a very complex device about which entire books have been written. If you're interested in learning more about a particular subject, volumes of information can be found in other texts and in manufacturers' literature.

Radio is a word used to define a system involving the transmission and reception of an electromagnetic wave upon which has been impressed some form of information. It can support one-way communication like AM or FM radio and television, also known as broadcast, where a single high power transmitter communicates with a large number of receivers. It can also support two-way communication as it does in cell phones, business band radios, and walkie-talkies, where each device contains both a transmitter and receiver (a transceiver). The common elements of these systems are that they require a transmitter to generate and impress information on the radio wave and a receiver capable of "hearing" the transmitted signal and returning it to its original state.

RF Energy

Radio Frequency energy (commonly abbreviated to RF) can simply be defined as an Alternating Current (AC) signal that forms a moving field of electric and magnetic force. These fields give rise to an energy field that propagates across space. Within this field the magnetic lines of force are always at right angles to the electric lines of force, and both lines of force are perpendicular to the direction of travel. The wave can have any position with respect to the earth over which it travels, and the plane in which the wave travels is called the *wave front*. Figure 2-1 illustrates the relationship of the magnetic field, the electrical field, and the resultant RF wave.

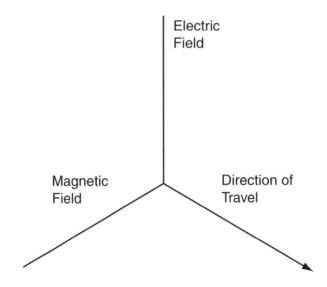

Figure 2-1: Lines of force

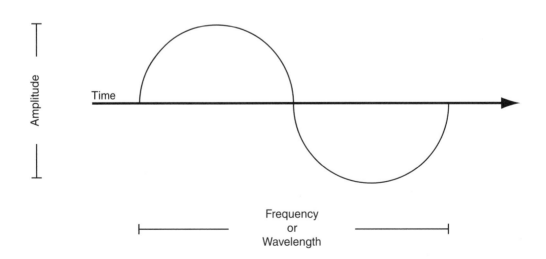

Figure 2-2: Sine wave

The RF field is defined by the characteristics of frequency (*f*) and wavelength (*w*). These two characteristics are inversely proportional to each other. Frequency is measured in units called *hertz*, after one of the founding fathers of RF, Heinrich Hertz. One hertz can be defined as one cycle of a complete sine wave per second. Wavelength is commonly measured in meters, and is defined as the length of the sine wave. In the atmosphere or in space the relationship is:

$$w = 300/f$$

where *f* is frequency in megahertz and *w* is wavelength in meters

Figure 2-2 shows a single cycle of a simple sine wave. The first thing you will notice is that this is an analog waveform. The amplitude of the wave is signified by its maximum height above the zero crossing, while the frequency is identified by the time it takes for the cycle to complete an entire positive and negative transition. If, for example Figure 2-1 showed a 1-Hz sine wave, the time from the initiation of the wave to the completion of the wave would be 1 second. A 10 Hz wave would complete in 1/10th of a second, and so on.

Much like visible light can be divided into colors based on wavelength, RF spectrum is divided into a number of frequency ranges, or bands, defined in Chart 2-1. This banding groups spectrum with common propagation and attenuation attributes together. As the frequency is increased beyond those associated with RF, the EM energy takes the form of infrared light, visible light, ultraviolet light, X-rays, and gamma rays.

Designation	Abbreviation	Frequencies	Free-space Wavelengths
Very Low Frequency	VLF	9 kHz – 30 kHz	33 km – 10 km
Low Frequency	LF	30 kHz – 300 kHz	10 km – 1 km
Medium Frequency	MF	300 kHz – 3 MHz	1 km – 100 m
High Frequency	HF	3 MHz – 30 MHz	100 m – 10 m
Very High Frequency	VHF	30 MHz – 300 MHz	10 m – 1 m
Ultra High Frequency	UHF	300 MHz – 3 GHz	1 m – 100 mm
Super High Frequency	SHF	3 GHz – 30 GHz	100 mm – 10 mm
Extremely High Frequency	EHF	30 GHz – 300 GHz	10 mm – 1 mm

Chart 2-1: Band allocation and frequency vs. wavelength relationships

The bands of frequencies are further divided into individual channels. These channels are nothing more than smaller slices of spectrum that are assigned to a transmitter and receiver and define their exact operating frequency. These channels are of varying spectral size and quantity, these characteristics being dictated by the band and type of communication service to be offered.

RF Generation and Transmission

Generation of the RF signal is the duty of the transmitter. The transmitter is comprised of a number of elements, each having a unique duty. A block diagram of a simple transmitter is shown in Figure 2-3.

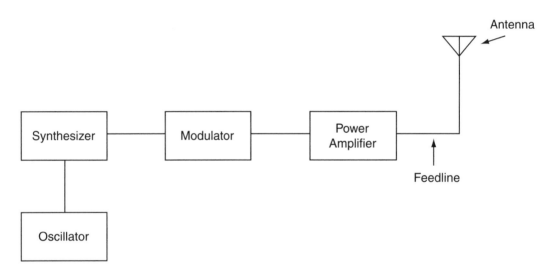

Figure 2-3: Transmitter block diagram

Oscillator

The first order of business is to generate the base RF signal at the desired operating frequency. This is the duty of the oscillator, or in modern radios, the frequency synthesizer. The oscillator functions by making use of the principles of amplification, feedback, and resonance. Most of us have heard the result of these principles at one time or another: the squeal that occurs when a microphone is too close to a speaker, such as occurs in a PA system on stage. This situation results in the formation of an unintended oscillator. An oscillator is an amplifier that has some of the output signal coupled back to the input in phase. This is known as positive feedback. If this coupling of input to output is allowed to continue, the amplifier will begin to stabilize around the resonant frequency of the components in the amplifier and the feedback loop. In the case of the PA system, the resonance is based upon the size of the coupling elements in the feedback loop: the speaker, microphone, and the air between them. In the same way a tuning fork generates a certain frequency based on its size, the physical size of the speaker, microphone, and air space between them determines the frequency at which the squeal (or tone) will be generated. This is known as resonance. Changing the dimensions of any of the elements will change the frequency of the tone. Since any frequency can be represented by a wavelength, the physical size of an object can determine the frequency at which it resonates. This concept will be explored further when we discuss antennas.

It just so happens that in the case of the PA system, the frequencies fall into the range of human hearing, and the medium conveying them (air) allows these mechanical vibrations to be heard. The same thing happens in the RF oscillator, but it normally happens at frequencies outside the range of human hearing and it occurs not with physical vibration of the air, but with electrical and magnetic vibration described by the AC signal.

Simple oscillators like those shown in Figure 2-4 rely on the resonant characteristics of either a tuned LC circuit or the resonance exhibited by a quartz crystal.

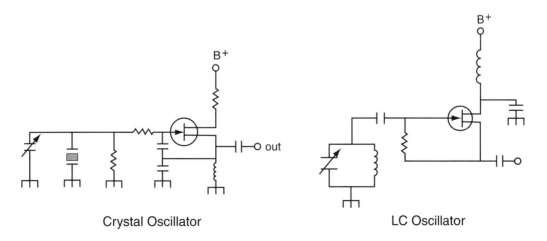

Crystal Oscillator   LC Oscillator

Figure 2-4: Simple oscillators

Inductors and capacitors exhibit a characteristic called *reactance*, which is frequency dependent and appears to the EM wave as resistance. Although reactance is measured in ohms just like resistance, unlike pure resistance, none of the power is dissipated by the reactance. Instead it is stored for a brief interval in the capacitor or inductor. Inductive reactance and capacitive reactance have opposing characteristics. Capacitive reactance affects low frequencies more than higher ones. In fact a capacitor will block DC (direct current) because it represents an infinite reactance to a non-AC signal. Inductive reactance, on the other hand, affects higher frequencies to a greater extent than lower ones. An inductor will pass DC as easily as a piece of wire.

When an inductor and capacitor are connected together, forming an LC circuit, an interesting interaction occurs. Since the reactances are working in opposition, there is a frequency where the capacitive reactance and inductive reactance are equal. This point of equality is also known as the resonant frequency. The resonant frequency can be determined mathematically as shown in Figure 2-5:

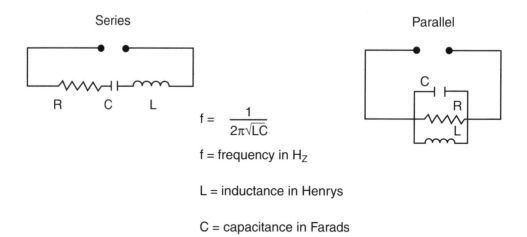

$$f = \frac{1}{2\pi\sqrt{LC}}$$

f = frequency in H_z

L = inductance in Henrys

C = capacitance in Farads

Figure 2-5: Series and parallel LC circuits and resonance formulae

Reactance and resonance are elements that are fundamental to radio and RF. They are used to characterize the behavior of all major circuits that make up a radio or RF circuit. LC circuits can be used as the basis for the resonant circuit in an oscillator, and by making the capacitor or inductor variable, some degree of tuning, or change in the resonant frequency is made possible. However LC circuits are not extremely stable. They tend to drift (change frequency) with temperature, and they are susceptible to outside influence, such as proximity to other objects. Due to these undesirable characteristics, LC circuits are not used as the frequency setting element of modern oscillators or frequency synthesizers. They are instead used where their broad response is of benefit: in filters and the tuned circuits associated with RF amplifiers. Instead of using an LC circuit as a frequency generating element, a more stable frequency generating element, the quartz crystal, is used.

Quartz is a substance that exhibits piezoelectric properties. Piezoelectric materials can generate an electric field when mechanically deformed. Conversely, they also deform when an electric charge is applied to them. So a resonator can also be constructed from a piece of quartz crystal sandwiched between two conducting plates. When voltage is applied to the conducting plates, it causes mechanical stress in the crystal. A frequency exists at which the crystal will start to vibrate, stressing and relaxing in a rhythmic pattern. The resonant frequency of a quartz crystal is dependent upon its physical size. The most important dimension is its thickness. Quartz crystals exhibit a very stable resonant frequency, and are commonly used in modern oscillators as to generate the fundamental frequency that will ultimately become the transmitted RF signal.

Unless the frequency to be generated is very low, say 100 MHz or lower, the oscillator cannot generate this frequency directly. This is due to the size of the crystal necessary to generate higher frequencies. As the frequency increases the size of the crystal becomes smaller. The energy dissipated by the crystal generates some amount of heat, and overheating can easily damage the smaller crystal. As well, as the heat generated causes fluctuation of the resonant frequency. Neither situation is acceptable, so the oscillator is run at the lowest possible frequency.

A problem with the simple oscillator is that it operates on a single frequency. Unfortunately modern communication devices require stable operation over a wide range of frequencies. In other words they need to be tunable. Doing this with multiple crystals is not effective, so modern radio equipment relies on a digitally generated and controlled oscillator called a *frequency synthesizer.*

Just as the name implies, this device synthesizes a sine wave of the appropriate frequency using digital logic to generate and control an analog signal. It uses a crystal controlled oscillator of a fixed frequency to provide a reference frequency that is used for both generating new frequencies as well as providing a reference to lock its operation to. Frequency synthesizers can be single IC devices, or they can be implemented along with the other radio functional blocks on a common chip. Because it is implemented in digital logic there is practically no limit to the granularity of the frequencies it can generate in its useable tuning range. This makes it an ideal building block for the modern radio, and without it modern modulation schemes like CDMA would be impractical or impossible to implement cost effectively.

Using a synthesizer removes many design complexities and substantially reduces the cost that would otherwise be associated with other forms of frequency generation. It also allows for the generation of the extremely complex modulation waveforms common in today's consumer radio hardware. Additionally, simply reprogramming its microprocessor controller can change its operating frequency and characteristics without any design change to the synthesizer itself.

Power Amplifiers

The next block in the simple transmitter is the RF power amplifier. The job of this amplifier is the same as any amplifier: faithfully to create a higher power image of the signal presented to its input. The amplifier takes the low power presented by the oscillator and increases the power to a level that will be sufficient to transmit the radio energy across the path between the transmitter and receiver. Three important considerations in power amplification are *power output, linearity,* and *efficiency.* Power output is measured in watts. Linearity is defined as the operating parameters of the device that result in a linear gain relationship between the input and output. Efficiency is the ratio of power output to total power input (the wattage demanded of the power supply). This value is typically expressed as a percentage, and is always less than 100%.

The linearity of Power Amplifiers is often described by the "class" to which the amplifier belongs. Class A, AB, B, C, and D are the common amplifier class descriptors. In Class A amplifiers, the amplifier is biased so that it is constantly conducting. These amplifiers are the most linear and the least efficient, as low as 10% efficiency is not unheard of. Class AB and B are biased so they do not constantly conduct. These amplifiers trade off some amount of linearity for increased efficiency. These amplifiers can achieve 35% efficiency. Class C amplifiers are biased so they do not conduct unless a signal of sufficient magnitude is input. The Class C amp is a non linear amplifier, and is useable only for those certain modulation types, like CW and FM, whose waveform does not require linear amplification to maintain modulation integrity. Class C amplifiers can achieve over 70% efficiency.

A relative newcomer is Class D amplification. This amplifier functions using pulse width modulation, where the desired signal is mixed with a triangle or square wave signal that has a significantly greater base frequency. The resulting output is a series of on-off pulses of varying width which correspond to the input signal. These pulses drive the amplifier output transistors. Since the pulses are either on or off, the transistors behave as they do in Class C operation. Class D amplifiers are much more complex than their traditional counterparts, but can achieve very high efficiencies and achieve extraordinary linearity at the same time.

The output of the power amplifier is sent to the next stage of the transmitter, the feedline and antenna. Antennas are resonant transducers used to convert between EM waves and AC signals in much the same way a microphone and speaker convert sound vibrations into AC signals and back again.

In the transmitter, the AC signal that has been amplified by the power amplifier is applied to the antenna. The antenna converts this AC signal into a varying EM field that can be coupled into air or space.

Antennas and Feedlines

The antenna is normally coupled to the transmitter or receiver by way of a coaxial cable, often just called *coax*. Coax is more than just a random piece of wire. It consists of four elements, each of which has an effect on its characteristic impedance and its loss. The four elements are the center conductor, the dielectric, the outer conductor and shield, and finally the outer jacket. The impedance of the coax is determined by the interaction of these elements. The diameter of the center conductor, the composition of the dielectric and its diameter, the diameter and construction of the shield and to a lesser extent, the material making up the outer jacket all effect the impedance and the loss of the cable.

In radio work, the most common coax has 50 or 75 ohm impedance. The most common for the systems discussed in this book are the 50 ohm varieties. Selecting the proper coax is important. The impedance must be correct in order for it to mate with the impedance of the antenna and radio. Energy flows with the least impairment when the impedances of all elements are equal. Using a cable or antenna with a different impedance from the radio causes an impedance mismatch. This mismatch causes energy to reflect away from the load back towards its source. This effect is also known as VSWR, or voltage standing wave ratio, which describes the match of the circuit and the amount of reflected energy. These reflected currents are dissipated as heat. Unfortunately heat is not RF energy, so every bit of energy that gets reflected is wasted.

Besides selecting the appropriate impedance, you also want to use the coax with the lowest loss. Unfortunately the lower the loss, the larger the diameter of the cable. Small cable such as RG174 is less than 3/16 inch in diameter and very flexible. At 2 GHz it exhibits 45 dB of loss per one hundred feet. LM400 cable is over ½ inch in diameter and exhibits 6.8 dB per hundred feet. It is also quite stiff and difficult to bend. Larger size coax can exhibit even less loss, although for a particular run you reach a point of diminishing returns as the size, cost, and installation complexity grows beyond the minor gains in loss performance.

Antennas come in a variety of shapes and sizes. Within any particular frequency band a multitude of different antennas exist. Each antenna has particular attributes that will make it suitable for specific purposes. **In general, antenna selection is one of the most important considerations when implementing a wireless data system.**

The perfect antenna is known as an isotropic radiator (Figure 2-6). It generates a perfect sphere of energy around it that has equal intensity in all directions. Isotropic antennas exist in theory but not in the real world. The closest visual analogy to an isotropic antenna would be the sun.

Assuming all the transmitter's power is emitted by this antenna, the amount of power flowing throughout an arbitrarily sized sphere will be the same as the energy emitted by the antenna. Thus the average power density measured in W/M^2, at the edge of the sphere can be defined as the ratio of total power and the sphere's surface area.

The simplest real-world antenna is the dipole. The dipole can consist of nothing more than two equal length pieces of wire that are of a length that is resonant at the desired frequency. The pattern generated by a dipole is in the shape of a torus, or donut. Now, if the same amount of energy is emitted by the dipole antenna, the average power density in a torus of equal diameter will be 2.1 dB higher, because the torus has less surface area than a sphere.

This results in the antenna acting like an amplifier, and apparently exhibiting gain! The same amount of electrical energy went into each antenna, but the strength of the resulting field at some distance from the antenna is greater with a dipole than with an isotrope, just because the energy is spread over a smaller area.

Two common terms you'll hear in specifying the gain of an antenna are dBi and dBd. These terms are defined as dB of gain referenced to an Isotrope (dBi) and dB of gain referenced to a dipole (dBd). A dipole has 2.1 dBi of gain, but 0 dBd of gain. It is critical to know which reference the antenna manufacturer uses, because it will have an effect on the path loss and power calculations we will discuss later.

Dipoles and isotropes are omnidirectional antennas, meaning they generate a field 360 degrees around them. A third type of antenna is the directional antenna. This antenna is designed in such a way as to concentrate its radiation in one direction, in a pattern that can be imagined as a cone. Think of a flashlight for a visual analogy. Once again, these antennas have gain, because you are concentrating the same energy over a smaller area. Figure 2-6 illustrates this concept using common light sources for analogies of antenna patterns.

Antennas have patterns defined in two planes: horizontal and vertical. Omnidirectional antennas are available with gains from 2.1 dBi to over 15 dBi. Since the horizontal pattern of the antenna is already defined as omnidirectional, the way the designer achieves gain in these antennas is by reducing the vertical size of the torus, or main lobe of the antenna. The vertical size of the main lobe of the

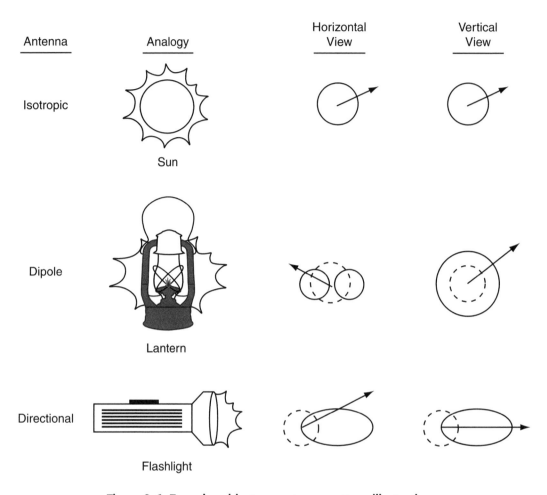

Antenna	Analogy	Horizontal View	Vertical View
Isotropic	Sun		
Dipole	Lantern		
Directional	Flashlight		

Figure 2-6: Everyday objects as antenna pattern illustrations

antenna is what is used to define the antenna's gain. The main lobe (also known as main beam) is defined in terms of its 3 dB or half-power points, that is, the points defined above and below the center of the main beam where the power falls by 3 dB.

As you can see in the antenna patterns in Figure 2-7, as the gain goes up the vertical beamwidth gets narrower. In very high gain antennas this beam can be reduced to less than ten degrees.

Now, this is a potential problem. If an antenna with a very narrow vertical beam is installed at a significant elevation above the area to be covered, there is a good chance that the majority of the gain will be wasted. That's because the power in

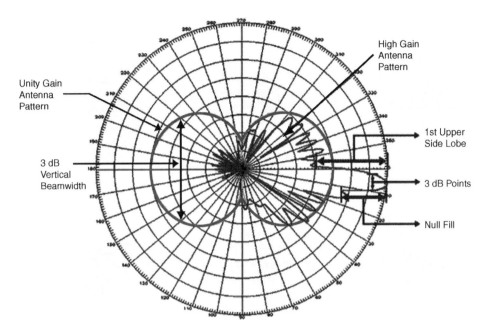

Figure 2-7: Comparison of the vertical pattern of unity gain (light grey) and high gain (dark grey) Omni antennas

the main lobe is pointing at the horizon, not at the receivers at ground level around the antenna. A sub-lobe having significantly less power than the main lobe is serving any receiver close by.

Not only is the power of the main lobe wasted, it can become an interferer to other users of the same frequency.

Antenna designers have a way to correct this situation, called *electrical downtilt*. In these antennas the main lobe does not point at the horizon. Instead the main beam is "tilted" towards the earth by a few degrees. This helps to put the power where it's useable, closer to the ground. Determining the amount of downtilt that can be used is a basic geometry calculation. Determine the area to be covered, and select an antenna height and downtilt that result in the main beam making ground-fall within the outer 25% of the coverage area. This concept is further explained in Chapter 5.

Directional antennas are available in many patterns as well. It is easier to get high gains in a directional antenna because it emits energy over less than 360 degrees. Common horizontal patterns are 180, 120, 100, 90, 60, 45, 30, and 15 degrees.

These antennas are useful in providing controlled area coverage. Still others are designed to support point-to-point links. These antennas are made with a parabolic reflector that focuses the energy into a beam that can be as narrow as 2 degrees. As shown in Figure 2-8, as compared to a 120-degree antenna commonly used to provide wide area coverage, parabolic reflector antennas have too narrow a pattern for area coverage, but because of their high gain can be used to extend the range of a system to a fixed location outside the nominal coverage area provided by the wider beamwidth antenna. Selecting an antenna that has appropriate trade-offs in gain and aperture is critical to ensuring that a radio-based system provides appropriate coverage to a designated area without generating significant amounts of interference in undesired areas.

Antenna behavior is reciprocal, meaning they behave the same while transmitting or receiving. So the first block in a receiver is the antenna we just discussed.

Figure 2-8a: Dish Antenna. Photo(s) and illustration(s) reprinted with the permission of MAXRAD, Inc.

Figure 2-8b: Sector Antenna. Photo(s) and illustration(s) reprinted with the permission of MAXRAD, Inc.

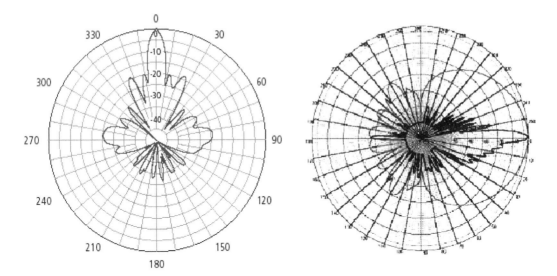

Figure 2-8c: Antenna pattern of a microwave parabolic dish antenna

Figure 2-8d: 120-degree sector antenna showing vertical (dark) and horizontal (light) patterns

RF Reception

In this case, the EM field passing across the antenna is transformed into an AC signal that is applied to the RF amplifier.

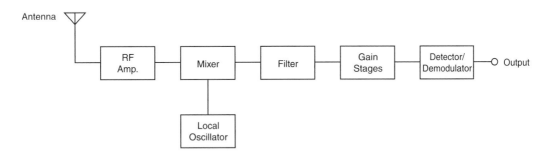

Figure 2-9: Receiver block diagram

By the time the signal gets to the receive antenna it has been attenuated by its travel through the environment. The signal at the antenna is also polluted with RF from a multitude of other sources. The first order of business is to put the received signal through a bandpass filter. The job of this filter is to remove any out of band energy so the RF amplifier is only presented with signals associated with the desired band.

The RF amplifier is a weak-signal amplifier designed to deal with exceedingly small input signals, in some cases measuring only a few picowatts (−100 dBm). dBm is a common term used to measure the power of radio signals. It is merely a reference of dB change referenced to one milliwatt. Figure 2-10 can be a useful reference when working with radio systems. In addition to illustrating just how tiny a signal is dealt with by the receiver, it can be used to quickly convert between watts and dBm. There is also a utility on the CD-ROM that allows you to accurately convert any value of watts to dBm or dBm to watts.

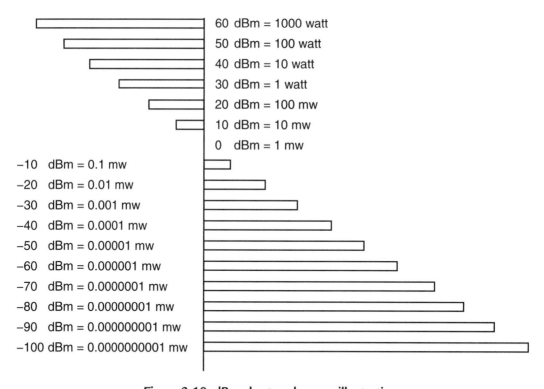

Figure 2-10: dBm chart and power illustration

Because of the vanishingly low signal power received by the antenna, the RF amplifier must generate minimal noise while increasing the signal voltage by a large factor. The specifications that denote the effectiveness of a weak-signal amplifier are *sensitivity*, defined as the ability to discern a weak signal, and the *intercept point* which defines how the device behaves in the presence of large signals. Design of an RF amplifier is always a trade-off between these two factors. Very low noise, high sensitivity amplifiers work well over a narrow range of input signal strengths. If the signal on the input gets too large, the device gets overdriven and no longer amplifies in a linear fashion. This is why filtering the signal coming from the antenna is critical. The energy contained in out of band signals can add to the total signal seen by the RF amplifier and cause it to be overdriven, even if the desired signal is of low intensity. On the other hand, devices that can tolerate high power signals exhibit higher noise and lower sensitivity. Selection of the best compromise is up to the designer of the receiver and is based upon the average signal levels that will be seen at the device's input.

The RF amplifier increases the level of the signal to a point where it can be processed by the mixer. The mixer gets one input from the RF amp and another input from a Local Oscillator (LO). The mixer combines these two signals, which results in four frequencies at its output: the receive frequency, the local oscillator frequency, and their sum and difference. For example assume we have a receive frequency of 2,400 MHz, and a LO frequency of 2,000 MHz. The mixer output would contain signals at 2,400 MHz, 2,000 MHz, 4,400 MHz and 400 MHz. The only valuable output is the difference signal at 400 MHz. This low frequency signal is easier to amplify and process, and is one of the reasons for using the mixer to generate it. The other reason is that by tuning the LO, one can tune the receiver to different frequencies, just like tuning the oscillator in the transmitter allowed it to generate different frequencies. The mix of signals goes through a filter that is resonant at just the difference frequency, allowing only it to pass. This filtered frequency enters a series of amplifiers called *Intermediate Frequency (IF)* amplifiers. The IF amp is tuned to the difference signal and provides additional amplification to the still miniscule signal. After the signal has been amplified enough to process it, it goes to a detector/demodulator. The detector is used to strip the intelligence from the signal and present that intelligence in its original form.

Now that we have a transmitter and a receiver, we have a communication system. The transmitter is capable of generating a carrier wave at the desired frequency. If coupled to an antenna, a signal would be radiated that could be detected by the receiver. Unfortunately, the carrier wave contains no information. It just exists as a steady state signal.

Modulation of RF Signals

In order to transmit information, something must be done with the signal created by the transmitter. That's where the other block in the transmitter, the modulator, enters the picture. The job of the modulator is to impress information on the carrier wave. It is possible to convey information by modulating any property of the carrier: time, frequency, amplitude and phase. Thus the simplest modulator is an on/off switch. If the carrier is turned on and off in a structured way following a known pattern, like Morse code for example, intelligence can be conveyed to the receiver. In its time, Morse code was a useful way to communicate; however, it soon fell to the wayside as other more useful modulators and modulation types became available for impressing first analog then digital information on the carrier.

An interesting thing happens to the carrier when it is modulated. No longer is the carrier the only component of the waveform. It now contains energy that occupies spectrum on both sides of the unmodulated carrier wave. This additional occupied spectrum is known as sidebands. So by addition of modulation to the carrier wave that previously occupied only a single frequency causes it to "spread out" and occupy more spectrum. The size and shape of the sidebands is dependent on the modulation, and can be determined mathematically.

This sideband energy defines the shape of the transmitted waveform, and is the reason why communication channels are assigned containing a specified amount of bandwidth. Rules and regulations governing spectrum allocation and use take into account the types on modulation to be used and the maximum amount of information that the modulation needs to represent, then divide the RF bands into channels that are sized according to the use of the spectrum.

Amplitude Modulation

Beyond just turning the carrier on and off, the other properties of the carrier that can be manipulated are its amplitude, frequency and phase. If you change the amplitude, or power contained in the carrier, based on the information to be conveyed, the resulting signal is known as Amplitude Modulation (AM). This is the modulation technique is illustrated in Figure 2-11, and is used as the modulation of choice in the AM broadcast band. Because AM is fairly easily accomplished, it was the first modulation method used after on/off switching. Because AM has the ability to impress a complex analog signal (like sound, for example) on the carrier it became immensely popular because it allowed the broadcast of voice and music. This modulation is still in worldwide use today by broadcasters. Even TV relies on a type of AM for transmitting the video portion of the TV signal.

AM, while simple, carries several penalties. First, it requires significant power change in the amplitude of the carrier. Second, because you are receiving a signal that varies in amplitude, any impulse noise that coexists in the channel, like lightning, auto ignitions, and fluorescent lights, is also heard by the receiver. You've no doubt experienced this firsthand when listening to an AM broadcast and hearing the impulse noise generated during a summer lightning storm.

In addition, an AM signal occupies more spectrum than a Morse code signal. The morse code signal occupies only the single frequency associated with the carrier wave. The AM signal spreads out from the carrier frequency to envelop spectrum equal to the carrier frequency ± the maximum modulated frequency. For example, if a carrier were generated at 10 MHz and a 3 KHz tone were AM modulated on to it, the result would be that spectrum from 9.007 MHz to 10.003 MHz, or 6 KHz of spectrum would be occupied by this signal. So more complex information can be sent at the cost of spectrum consumed. This is an artifact of any modulation, and is one of the factors that determine the size, in Hz of bandwidth, of the channel.

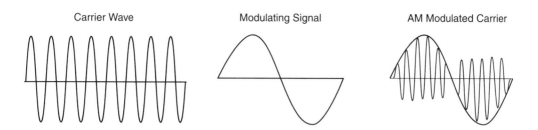

Carrier Wave   Modulating Signal   AM Modulated Carrier

Figure 2-11: Illustration of amplitude modulation

Frequency Modulation

The next evolution of modulation was Frequency Modulation (FM). As illustrated in Figure 2-12, FM works by manipulating the frequency of the carrier in concert with the incoming information. Once again, the first consumer use of FM was in the broadcast arena. FM is still used today in the FM broadcast band, and is used to carry the audio content in a TV signal.

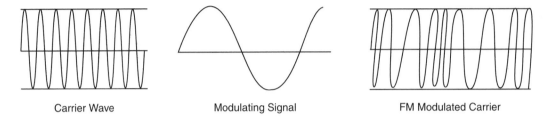

| Carrier Wave | Modulating Signal | FM Modulated Carrier |

Figure 2-12: Illustration of frequency modulation

A variation of FM is Phase Modulation (PM). This modulation method is tightly coupled to FM because you cannot change the phase of the carrier without also changing its frequency. FM and PM require little energy to accomplish and since the receiver is looking only at frequency or phase shifts, this modulation technique is also very noise tolerant since you are interpreting frequency changes and ignoring amplitude variations.

From a spectral use standpoint, FM is similar to AM in that it occupies spectrum around the main carrier. It can at minimum occupy similar bandwidth to an AM carrier, but it is possible to configure the channel to occupy a great deal more spectrum in order to increase the maximum modulated frequency and maximize the ability for the transmission to be received effectively. The actual spectral occupancy of an FM signal is a complex calculation dependent on the channel bandwidth, the modulation index, and the maximum frequency to be modulated.

This artifact of modulation increasing the size of the occupied bandwidth has given rise to the concept of channelization of RF spectrum. The act of impressing information on a carrier causes it to "spread out" and occupy a mathematically calculable amount of spectrum on either side of the carrier's center frequency. In order to avoid random interference between users of spectrum, the world's RF management bodies sliced up, or "channelized" the spectral allocations in accordance with the modulation to be used in the band. Regardless of the band, if you look at the rules regulating its use, there will be specific channel masks that specify the spectrum that can legally be occupied by a transmitted signal.

More complex modulation can carry more information, so the channel allocations are larger. Case in point: an AM radio broadcast station is allocated a channel 10 KHz wide. This channel can carry audio information of ± 5 KHz, and is thus wide enough to carry voice and low fidelity music. An FM radio station is allocated a 200 KHz channel which can carry an audio signal of up to 15 KHz. A TV station, because of the complexity of the information broadcast, needs 6 MHz of spectrum to contain the modulated signal, and an 802.11b carrier occupies in excess of 20 MHz of spectrum!

Complex Modulation

AM, FM, and PM provided simple ways to convey the only information available at the time of their invention: audio. As digital information became available, these modulation techniques were pressed into service to transmit digital information too. As seen previously in this chapter, an RF carrier is a sine wave, and a sine wave is analog in nature. To transmit digital information required the digital information to be converted to the analog realm. This was accomplished by use of an analog MODEM, or **MO**dulator/**DEM**odulator, whose sole purpose was to change digital information into audio frequency tones that could be transmitted and received over a radio carrier or phone line. Originally, modems worked by generating two distinct audio frequencies. Each tone was associated with either a binary 1 or 0. As the binary information entered the modem serially, it was converted to audio tones to be transmitted.

As the amount of data needing to be transmitted increased, the simple two-tone modem became incapable of meeting the ever increasing throughput requirements. Fortunately, development of the digital logic and processing power responsible for this growing bandwidth requirement also gave rise to the ability to manipulate or modulate the carrier wave in ever more creative and complex ways. Furthermore, since we can now use this same processing power to transform complex analog signals like video and voice into digital form, there is no longer a need to have the modulation support these complex analog waveforms that gave rise to AM and FM. Instead, we can modulate the carrier in such a way as to represent only bits and bytes.

This is a paradigm-changing concept. No longer does the waveform need to represent complex analog information. It merely needs to convey bits and bytes, which can be conveyed as "states" of the carrier wave. Take for example AM; the carrier states could be full power for a 1 and half power for a 0. FM could represent 0 and 1 based on the shift of the frequency. Freed from the need to support an analog waveform, designers began looking at how individual carrier states could be used to represent bits of digital information.

There are numerous digital modulation techniques in service today, but in one way or another, they all manipulate the same elements of the carrier wave: time, amplitude, frequency and phase. It's worth mentioning, that the universal principle of TINSTAAFL (there is no such thing as a free lunch) exists in radio too. Any selected modulation technique makes trade-offs between spectral occupancy, maximum information rate, circuit complexity, power requirements, and robustness of the signal.

Actually in radio the TINSTAAFL principle has a real name: Shannon's Information Theory, named after Claude Shannon. Shannon was a mathematician at Bell Labs. In 1948 he authored a Bell System Technical Journal article entitled "A Mathematical Theory of Communication" in which he postulated that, due to entropy, uncertainty was a fact of life in a communication channel. Thus there was no reason why more information could not be transmitted in a given channel so long as you could tolerate the rising uncertainty, or error rate, in the received information. In other words, simple modulation is very robust, but with robustness comes severe limitations on capacity. Complex systems are less robust, but have more potential capacity. Complex systems also need more power density at the receiver in order to increase the certainty of the transmitted state. The terms C/I, C/N, and Eb/No are commonly associated with communication systems. These terms mean Carrier to Noise or Interference ratio (C/I, C/N, C/I+N) and Energy per Bit relative to Noise ratio (Eb/No). These terms are used to identify how strong the signal must be in comparison to the noise and interference in the channel in order for the receiver to reconstruct the transmitted information accurately.

The equipment's performance will be compromised if the signal is allowed to fall below the specified threshold. Meeting this criterion becomes a key requirement of system design and implementation. This value, along with receive sensitivity and transmit power, define the limit of the reliable distance over which the system can communicate in a given environment.

When information is to be transmitted and received, there are a complex set of variables that govern the transaction. These include: the bandwidth of the digital signal, its transfer rate, size of the transmission channel, noise in the transmission channel, interference in the transmission channel, complexity of the modulation, propagation delay, reliability of the transmission channel, transmit power, receive sensitivity, error coding and error correction algorithms. These variables all have effects on the accuracy and error rate of the recovered signal. One of the most important applications of Shannon's theory is using it to determine the appropriate trade-off among the competing variables. This allows the designer to optimize the system to fit a given set of parameters.

Modern digital communication equipment have throughput requirements so high, and spectrum has become so congested, that advanced modulation techniques are needed in order to achieve the desired throughput in narrow channels. The modulation formats still use phase and amplitude as the modulated characteristics, but have implemented them in ever more complex ways in order to increase the throughput of the channel.

Since digital radios are no longer dealing with analog information, they do not have to be based on modulations that support analog signals. They merely have to transit 1's and 0's. This can be done simply with two phase or amplitude states: one state representing a binary 1, the other representing a 0. In order to transmit data faster, you need more transitions. Luckily, because of the number of discrete phase angles available (theoretically 360 but practically far less) and the number of amplitude states available (theoretically infinite, but again practically far less), a carrier transition can represent more than one bit. If four distinct carrier states are available, 2 bits can be represented by each transition, eight states yield 3 bits, and so forth.

BiPhase Shift Keying (BPSK) phase modulates the carrier with two distinct phase shifts, 180 degrees opposed, it can represent 1 bit per transition. Figure 2-13 shows this concept using signal states at 0 and 180 degrees. There is no reason why the initial phase state must be 0. So long as the phase states are 180 degrees out of phase, any two states could be used. 45 and 225 degrees are common states used in BPSK radio equipment.

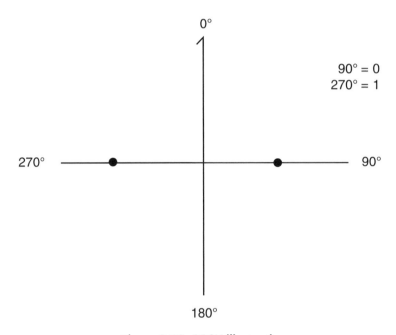

Figure 2-13: BPSK illustration

If two phase states can be used, why not more? According to Shannon, it's possible if you are willing to use more power to communicate over the same distance. Quadrature Phase Shift Keying (QPSK) is the next logical step up the modulation complexity curve. As shown in Figure 2-14, QPSK uses four distinct phases, each separated by 90 degrees. It can represent two bits per transition, but in return requires more signal power at the receiver in order to recover the transmitted information accurately. As previously discussed, there is no reason why one state must be 0 degrees. As in BPSK, an initial state of 45 degrees is commonly used.

Further increases in efficiency can come from adding even more phase states. Doubling the number of phase states in QPSK yields 8 PSK, which uses eight distinct phases separated by 45 degrees, and can represent three bits per transition.

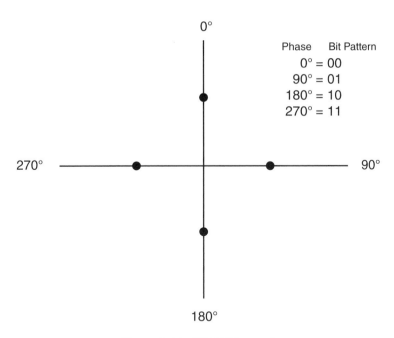

Figure 2-14: QPSK illustration

Each of these modulations is more efficient in that more bits are transmitted each second. At the same time, as Shannon predicted, each is more susceptible to loss of information in a noisy transmission medium. In a noisy and constantly changing medium like radio, it's relatively easy to pick out the two phase states of BPSK, but much harder to determine the 45 degree shifts of 8 PSK accurately. Further additions of phase states continue to reduce the robustness of the radio channel. In common practice 8 PSK is the highest order PSK modulation in use.

As spectrum became more congested and the amount of information increased, even 8PSK proved insufficient to provide enough channel capacity in many cases. This limitation was overcome by using the carrier's amplitude to convey additional bits. So in addition to modulating the phase, engineers started modulating the amplitude as well. This is known as QAM, or Quadrature Amplitude Modulation and is a fancy name for a simple process. If you take the two phase states of BPSK, and add two distinct amplitude states to each, you have QAM. This concept is illustrated in Figure 2-15. The signal is a basic BPSK signal with 0 and 180 degree phase states, but now each phase state is also transmitted with two unique amplitudes.

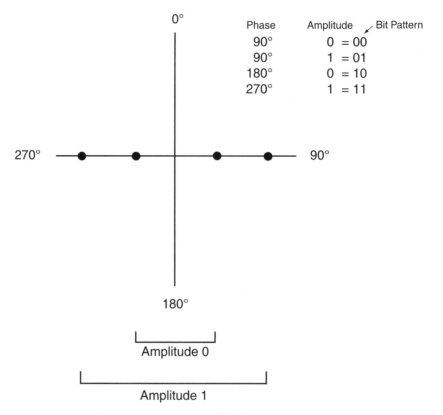

Figure 2-15: QAM Modulation

Of course it doesn't stop there. By adding two distinct amplitude shifts to a QPSK signal you get 8 QAM, which has eight distinct phase/amplitude states. Each of these states can represent 3 bits per transition, the same as for 8 PSK.

But wait, there's still more! 16 QAM has four phase states with four amplitude states, can represent 4 bits per transition, 32 QAM can represent 5 bits, 64 QAM can represent 6 bits, and 256 QAM, which has sixteen phase states and sixteen unique amplitude states, can represent 8 bits per transition. All of these modulations are commonly used in modern equipment. In fact some modern point-to-point microwave equipment uses 512 QAM.

As you can imagine, the uncertainty associated with receiving and correctly interpreting one unique state out of 256 is extremely high. In fact, the carrier must be at least 30 dB, or 1000 times stronger than the noise in the channel, in order to be heard and correctly demodulated by the receiver. Modulations this complex can only operate over the cleanest mediums, and even then require significantly more power than less complex modulations. Fixed microwave links and communication over coaxial cables, like cable modems, can use this modulation because they are very noise free and fade free transmission mediums. BPSK on the other hand needs to be only 6 dB, or four times stronger than the noise in the channel. For this reason, designers of mobile communications systems rely on the simplest modulation that will get the job done. The balancing act here is power versus spectrum bandwidth for a given throughput. Simple modulations require lower power to cover a given area effectively, however they provide limited throughput. On the other hand, complex modulation requires more power to cover the same area, but offers increased throughput in a given channel. If too complex a modulation is selected, the system may require too much power of a portable device, leading to short battery life. Or the system may have severe coverage limits, or in the worst case, the system may be fragile and experience so many errors that effective communication is impossible over the desired coverage area of each transmitter. Figure 2-16 illustrates the characteristics and trade-offs associated with increasing the modulation complexity.

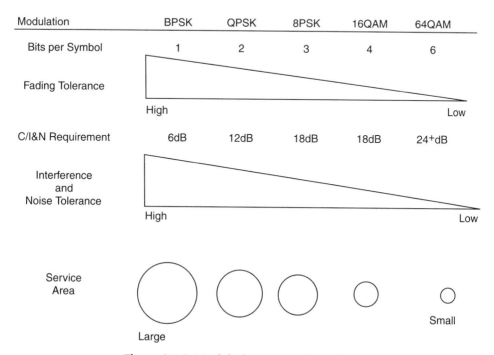

Figure 2-16: Modulation comparison diagram

Unlike a voice communication system, a data communication system cannot tolerate errors. If the transmitted signal is not received 100% correctly, the transmission is useless, and the information must be retransmitted. Because of the uncertainty associated with the reception of the signal, all digital communication systems have implemented error checking and error correction algorithms. These algorithms help to identify, and to some extent correct, errors caused in the transmission and reception of the signal. This allows the channel to tolerate some level of error and allow corrective action to be taken. These algorithms allow the signal to noise ratio to become worse than the theoretical requirements while still providing a useful communication channel. Again, there is a trade-off. These algorithms add overhead to the transmitted information, because they add either extra or redundant information to the desired data so that the received signal can be recreated correctly. Error correction algorithms use information coding and redundancy schemes to send enough extra data to allow for the fact that some received states will be indeterminate. By using the redundant data, the original information can be recreated if enough good data is received. Error checking algorithms are used to test for errors. If errors are detected, the error correcting algorithms are brought into play. If they cannot correct the data, then the system asks the transmitter to resend the data.

Duplexing

So far we've discussed a transmitter and receiver as standalone devices capable of a one-way communication. Wireless data systems must be two-way, or duplex systems, if only to allow the receive end to acknowledge receipt of good information, or ask for a resend of errored information. There are two forms of duplexing available: Frequency Division Duplexing, and Time Division Duplexing.

Frequency Division Duplexing

Frequency Division Duplexing is accomplished by allocating two equal, distinct, and separate frequencies to the communication channel. One of the frequencies is transmitted by the base station transmitter and received by the remote station and the other frequency is transmitted by the remote station and received by the base station. Since these duplex systems share a common antenna, the two frequencies assigned have large separations between them, 45 MHz or more, in order to assure that the local transmitter energy can be easily filtered out of the local receiver. Cellular and PCS phones use FDD. FDD is most useful in systems expecting symmetric traffic, because the two channels assigned are equal in bandwidth.

Time Division Duplexing

In Time Division Duplex (TDD) allows the use of a single frequency to accommodate both transmit and receive duties at both ends of the link. This is accomplished by timeslicing the channel fast enough so the transmitters and receivers see a continuous flow of information. The channel is temporally divided into transmit timeslots and receive timeslots with a small guard time between them. 802.11 equipment and some cordless phones working in the 2.4 GHz-band use TDD. TDD is useful in systems that have asymmetric traffic patterns, because the time slots can be allocated asymmetrically.

Regardless of the duplex method used, both the base station and the remote user need to have both a transmitter and receiver, or transceiver. This is what you purchase when you buy wireless data radio equipment. All the blocks and capabilities discussed above have been considered, trade-offs have been made based upon the expected use of the equipment, and a product with some limited capability has been produced and marketed. By the nature of the trade-offs made in the design of the equipment, there is no "one size fits all" solution. As the constructor of a communications network, it is up to you to select an appropriate hardware solution that best meets the needs of your users. Cost, capacity, range, and reliability are some variables that you will have to consider in selecting an appropriate hardware solution to meet the users needs.

Multiple Access Techniques

The term users (plural) is an important distinction. The communication networks we are planning to implement are not designed to support the needs of a single user. If the network is effectively implemented and marketed to users, there will be multiple users simultaneously trying to use the capacity of the network. This brings us to the final topic to be discussed in this chapter: Multiple Access (MA) techniques.

With the advent of digital cell phones and PCS phones, we've all heard the terms TDMA and CDMA. These are two forms of Multiple Access technologies. Simply stated MA techniques allow for the sharing of spectrum and capacity by multiple users.

The original MA technique was Frequency Division Multiple Access (FDMA). As shown in Figure 2-17, FDMA is no more than dividing the RF band up into discrete channel allocations. Each of the channels is assigned enough bandwidth to accommodate the modulation technique and information rate requirements of the technology that is to use the channels. Each channel is available for one user at a time. Broadcast radio and TV are an example of FDMA. Each broadcaster is allocated a channel, which is wide enough to contain all the information they are transmitting. This was also the technique used in early cellular phone systems. Each carrier was allocated over 300 30KHz wide channels that used analog FM modulation. The service provider made these channels available to on a shared use basis for users as they were needed to make or receive a phone call. When no one was using them they were in an idle channel state waiting to provide service.

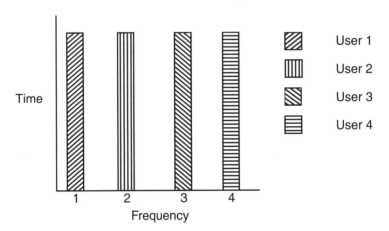

Figure 2-17: FDMA illustration

Another method of splitting up channels is shown in Figure 2-18. This method uses time slices to separate users. This is known as Time Division Multiple Access (TDMA). TDMA relies on the fact that if you can switch the connection back and forth between users fast enough, the user will never know they are sharing the channel. In TDMA, each channel is divided into a number of time slots. These

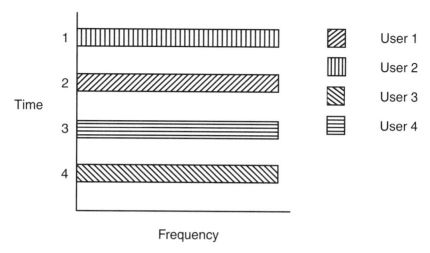

Figure 2-18: TDMA illustration

time slots occupy the entire channel, and are assigned to users instead of assigning the entire frequency all the time. Some cellular operators used TDMA to increase the capacity of their networks when FDMA could no longer provide economic growth. The 30 KHz channels were converted to digital channels containing three time slots. Now three users could simultaneously occupy the spectrum that was previously available to only one user. The need for more channels gave rise to replacing analog FM-based systems with digital-based systems and the inevitable trade-off of fidelity and simplicity for complexity and capacity.

FDMA and TDMA require careful frequency management and use planning because individual channels can be constantly in use. The use of individual channels leads to potential interference if different people in the same area simultaneously use the same channel. This results in the need to introduce a new system design criterion: frequency reuse planning. Reuse planning, as it's commonly called, is the science of managing the deployment of a limited amount of spectrum in order to yield the maximum capacity in an area to be served. As you can guess by the name, reuse is the deployment of the same channel multiple times in a single geographic area in such a way that interference among the co-channel reuse sites is minimized. This is accomplished by careful selection of site location, antenna selection, antenna height, and transmit power output. Careful consideration must be given to these factors in order to achieve interference free operation. Frequency reuse will be further discussed in Chapter 3.

Spread-Spectrum Modulation

FDMA and TDMA systems cannot easily accommodate the conflicting requirements of interference rejection, high throughput, and multiple users without active frequency management and interference control, so other modulation and multiple access technologies were perfected. Spread Spectrum is one of these technologies. Originally utilized as a method to provide secure communications, it requires a large bandwidth to function because, as the name implies, the carrier is spread out over a channel much larger than the fundamental bandwidth required by the transmission.

There are two variations of spread spectrum: Frequency Hopping (FHSS) and Direct Sequence (DSSS). They both work on the statistical principle that usage collisions are inevitable, but if the spreading or hopping codes are each unique, then there will never be cases where two or more users consistently collide on common channels. If collisions do occur, and data is lost, it can normally be recovered on the next reception interval, because statistically, there will not be serial collisions. Further, because each user has a unique spreading code, the receiver can use this code to discriminate the desired communication from all the others. This effect reduces the need for frequency management and interference control, and shifts the radio design to instead consider site coverage and capacity as the design variables. These two factors interrelate because while spread spectrum techniques can isolate individual users from each other, they still generate RF signals which raise the noise floor of the communication channel. As users occupy the channel, the noise floor rises. With increased noise comes the need for increased transmit power to overcome this noise. Essentially what happens is that the coverage area of a site becomes dependent on the number of active users on the channel. The maximum coverage of a site will be achieved with just a single user. As additional users are added to the system, each receiver must deal with the increased noise generated by other users. Since this new noise is additive to the thermal noise in the channel, it affects the ability of the receiver to receive and demodulate the transmitted signal accurately. If power is limited (as it always is in the real world), then the additional noise will cause the reliable coverage area of a site to shrink as users are added to the system.

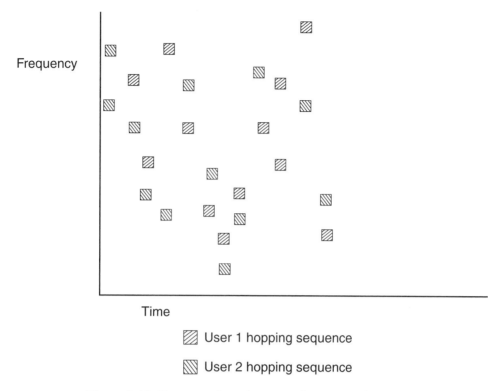

Figure 2-19: Frequency hopping spread spectrum (FHSS)

One spreading solution illustrated in Figure 2-19 is FHSS. FHSS relies on an FDMA channel set. The transmitter and receiver are frequency agile and have the ability to tune (hop) to any of the available channels many times a second. This allows the use of individual frequencies to be time spread, thus lowering the occupancy rate on any channel and lowering the average interference level on any one channel. Each user is assigned a unique hopping sequence that tells the transceiver the sequence of channels to use for its communication.

Another spreading technique is Direct Sequence Spread Spectrum (DSSS) in which a pseudo-random code is used to spread the carrier. As illustrated in Figure 2-20, the data to be communicated is XORed with this pseudo-random code. The resulting spread signal is used to modulate the carrier, resulting in a signal that is so widely spread that it becomes indistinguishable from thermal noise. In the receiver, the carrier is received and amplified. The code contained in the received signal is mixed with a local carrier to recover the spread digital signal. This

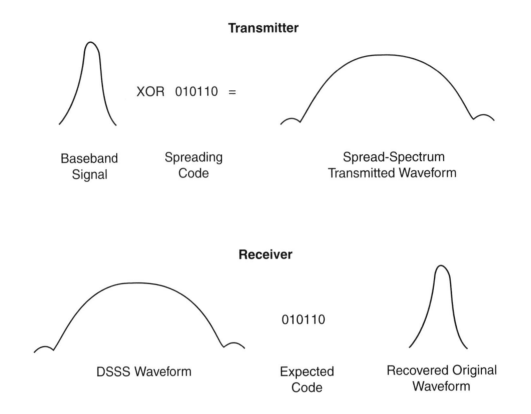

Transmitter

XOR 010110 =

Baseband Signal

Spreading Code

Spread-Spectrum Transmitted Waveform

Receiver

DSSS Waveform

010110

Expected Code

Recovered Original Waveform

Figure 2-20: Direct sequence spread spectrum

received code is locked to an internally generated pseudo-random code matching the anticipated signal generated by the receiver. The received signal is correlated with the self-generated code, thus extracting the information. By spreading the information over a wide channel, DSSS offers protection from narrow band interferers, because the narrow band interferer affects only a small portion of the overall spectrum assigned to the carrier. Because the energy density on any discrete frequency is miniscule, DSSS spread spectrum lends itself to implementation in systems that have many users and do not use interference management.

One implementation of DSSS is known as Code Division Multiple Access (CDMA). This is not a new modulation; instead it is a way of using the characteristics of DSSS to provide multiple users simultaneous access to the channel. In essence, CDMA is nothing more than a DSSS system that uses a number of unique orthogonal codes as the pseudo-random spreading codes. By associating a unique code with each user on the channel, the CDMA system can support multiple users on the same channel using the spreading code to isolate them from each other.

The unlicensed, or Part 15 bands, are required to accommodate many users of the same spectrum with no frequency coordination requirements. Further, the FCC originally mandated that the techniques used to modulate the carrier would trade off throughput for interference robustness. This mandate led to Spread Spectrum being adopted for the 802.11 specification. 802.11 supports 1 and 2 Mbps throughput in a 20 MHz wide channel. The reason for this poor efficiency was to assure that multiple uncoordinated users could co exist on the same channel without unmanageable interference. The rules were later relaxed, and allowed spectral efficiency to play a greater role in the trade-offs. This lead to the development of 802.11b/a/and g. These higher throughput standards provide more data bandwidth in the same spectrum, but are less tolerant of interference and noise.

DSSS is used in 802.11b when it is operating at the 1 and 2 Mbps rates. At these rates the spreading technique uses a single-spreading code, called the *Barker Code*, for all devices. At the 5.5 and 11 Mbps rates, the coding shifts to Complimentary Code Keying (CCK). CCK increases the throughput of the channel by using each of 64 orthogonal codes to represent six unique bit patterns. The final 2 bits that make up the 8-bit data byte are generated by using one of four phases to modulate the carrier. Unlike CDMA, where each user is assigned a unique code, CCK uses all the codes to represent data bits, thus the receiver cannot use the codes to discriminate one user from another.

To allow multiple users access to the available bandwidth, 802.11a/b and g use a sharing method called *CSMA/CA*, or Carrier Sense Multiple Access with Collision Avoidance. In this way, these standards work like traditional Ethernet: there is no central coordination of data broadcasts. Unlike Ethernet, which is a true full time duplex system, these wireless standards use TDD as their duplexing method, so there is no way to listen to the channel while transmitting. Thus there is no method of collision detection available. Instead, the best that can be done is collision avoidance. The device having data to send listens to the channel. If it senses that a carrier is there (the channel is in use) it will not transmit. Instead, it will be assigned a random retry interval, after which it attempts to retransmit. If it senses the channel is still in use, it backs off twice as long as the first random interval before trying again. If successful in transmitting, the station then awaits an ACK message. If an ACK is not received, the station assumes a collision occurred and attempts to resend the data during the next interval when it senses an idle channel.

CSMA/CA allows both multiple users on an individual Wireless LAN (WLAN) and permits multiple Wireless LANs to coexist on the same channel in proximity to each other. This is important because the 802.11 standards were designed for use in unlicensed spectrum, where there is no requirement for frequency usage coordination. This does not, however, mean that multiple co-channel facilities do not interfere with each other. In fact they do. Interference generated by different WLANs on a common channel results in all devices hearing the multiple carriers, and CSMA/CA will function to protect all LANs. This results in the multiple WLANs and their users sharing the throughput of the channel. In other words if two 802.11b WLANs were working in a common area sharing channel 6, neither WLAN could achieve full utilization of the available 11 Mbps. Assuming there was sufficient traffic to fill the available bandwidth, each LAN would nominally get half the available bandwidth, or 5.5 Mbps. This sharing will continue to happen if more WLANs using the same channel appear in the area.

OFDM

802.11a, 802.11g, 802.16 and 802.20 all rely on a relatively new modulation technique known as Orthogonal Frequency Division Modulation (OFDM). OFDM is a variation of FDM. Instead of using a single carrier within the channel, OFDM uses a large number of small overlapping channels (as seen in Figure 2-21) to transmit the data to be conveyed. Each of these sub-channels (also called "tones") has its own independent modem, and appears to be an independent carrier. These carriers overlap, but are spaced apart at precise frequencies so as to provide "orthogonality." The center of the modulated carrier is centered on the edge of the adjacent carriers. This technique prevents the independent demodulators from seeing frequencies other than their own. The benefits of OFDM are high spectral efficiency, great flexibility to conform to available channel bandwidth, and lower susceptibility to multipath distortion. This is useful because in a typical terrestrial

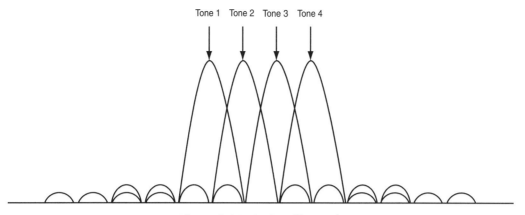

Figure 2-21: OFDM illustration

propagation environment, there are signal reflections (i.e. the transmitted signal arrives at the receiver from various paths of different length) that cause distortion of the received signal. As always, there is a trade off: OFDM is more susceptible to interference, especially from narrowband devices, and it requires extremely stable oscillators since it can tolerate little frequency drift. Once again this is a modulation technique that has been known for many years, but has only recently become feasible for consumer grade equipment due to the falling cost and rising complexity of digital circuitry and computing power.

In OFDM, each of the orthogonal carriers can be independently modulated with a BPSK or QAM signal. Because they are treated as independent channels, the selected modulation on each channel can be tailored to the fading environment of the propagation path. Implementing this flexibility adds complexity to the system, but in return allows the maximum throughput to be accomplished because it can dynamically accommodate the frequency selective fading of the channel. If a certain subcarrier occupies a faded frequency, it can be assigned a lower order modulation. If the subcarrier is unfaded, it can operate at the maximum modulation complexity.

Current consumer equipment implementations of OFDM like 802.11a and 802.11g do not implement this complexity. These standards use a common modulation on all subcarriers. Manufacturers of proprietary solutions offer varying levels of complexity based upon the anticipated use of the hardware. OFDM comes in many flavors, depending on the manufacturer and the intended use of the equipment. Just like all the other technologies we've discussed, the implementation trade-offs are selected by the standards body or equipment designer in order to maximize the equipment's utility in a given market space.

OFDM, because of its flexibility and high spectral efficiency is being considered as the technology for 4th generation cellular systems, and is being used in more and more standards-based and proprietary data communications products. It is also the basis for wired ADSL technology and some HDTV transmissions standards.

Ultra Wideband

Ultra Wideband (UWB) communications systems are also referred to as carrier free, baseband, or impulse technology. The basic concept is to generate, transmit and receive an extremely short duration burst of radio frequency (RF) energy. These "pulses" typically range from tens of picoseconds (trillionths of a second) to a few nanoseconds (billionths of a second) long. Each pulse, or burst, represents only one or two cycles of an RF carrier wave. The resulting waveforms are so extremely broad that it is often difficult to determine an actual RF center frequency of the transmission. The earliest methods of UWB signal generation used a fast rise-time pulse coupled to a wideband antenna to generate the RF signal.

In fact, the earliest radio signals generated by Hertz and Marconi could be called *ultra wideband* because they were generated in a very similar way and occupied vast amounts of spectrum. In the Hertz and Marconi transmitters, a spark was used to excite a tuned circuit and antenna, thus generating a very wideband signal that covered frequencies from tens of kilohertz to tens of megahertz. Such spark gap transmitters were the norm in the early days of radio and were used to transmit the digital code of the day: Morse code.

As radio advanced in complexity and utility, it began to control the carrier frequency more carefully in order to allow many users to use the technology. Control of operating frequency was needed in order to allow analog modulation like AM to be used, and to allow coexistence of the many voice-based radio systems that were coming into use. The spark gap transmitter with its wideband spectral output could not support analog modulation and could no longer be operated on an interference free basis because its output overlapped so many assigned channels.

Modern UWB systems take advantage of the extremely sensitive RF amplifier technology currently available, and therefore require substantially less transmit power than their spark gap predecessors. In fact, modern UWB signals are of such a low spectral density that the output of a single transmitter may be indistinguishable from the noise floor.

Since UWB waveforms are of such short time duration, they have the property of relative immunity to the multipath cancellation effects observed in mobile and in-building environments. As discussed in Chapter 3, multipath cancellation occurs when a strong reflected wave arrives partially or totally out of phase with the direct path signal, causing a signal cancellation at the receiver. Because of the very short pulses, the receiver can be timed to look for the direct path signal. Since the reflected signal arrived by way of a longer path, it takes more time to get to the receiver, thus it is time delayed in reference to the direct path. Therefore it is quite possible that the direct path signal has come and gone before the reflected path arrives; thus no cancellation can occur. This makes UWB systems well suited for nomadic or mobile wireless applications where constant multipath leads to fading and signal integrity issues.

Since the bandwidth of the generated signal is inversely related to pulse duration, the spectral bandwidth occupied by these waveforms can be made quite large. Therefore the resulting energy densities (transmitted watts of power per hertz of bandwidth) can be quite low. This low energy density translates into a low probability of interference to other services also occupying the band. This is not to say that they can operate on an interference free basis. The laws of physics still rule, and every UWB signal that is generated adds to the noise floor of the band in which it operates. Any rise in noise floor has a negative effect on the operation of other radio devices due to the need for more power at the receiver in order to overcome the increased noise floor.

One of the advantages of UWB technology is low system complexity and low cost. UWB systems are inherently digital because of their impulse nature, and can be constructed with minimal RF related electronics. Because of their inherent RF simplicity, these systems are quite frequency agile, thus enabling them to be easily designed to operate anywhere within the RF spectrum.

Due to the interference concerns surrounding UWB, the FCC has not authorized it for outdoor use, and many other countries have not authorized its use at all. Most of the focus for this technology is for deployment of a very high speed (>200 Mbps) networking technology to support short range communication (1 to 50 feet) for in-building or in-room networks. In 2003, the IEEE 802.15 committee was working on a standard for just such a solution. While many of the solutions presented to this committee are impulse-based, at least one of the solutions discussed was a wideband (>500 MHz) OFDM solution.

So, it remains to be seen whether impulse UWB becomes a pervasive standard for data networks. It has certain advantages that make it ideal for low cost high performance personal area networks, but it also has the possibility of degrading other communication systems if UWB equipment is sold and used in high volume. Even though this technology differs significantly from conventional RF devices, from a system design standpoint it still follows the basic laws of physics and the design guidelines and tools used for designing conventional RF networks can be used to design UWB systems as well.

CHAPTER 3

Propagation, Path Loss, Fading and Link Budgets

■ Path Loss and System Coverage

■ Frequency Reuse

Propagation, Path Loss, Fading and Link Budgets

Understanding how radio waves propagate through space is critical to the design of any radio-based network. Radio is an electromagnetic wave whose propagation is affected by many variables. Frequency, distance, terrain, objects in the wave's path, and reflections all have an effect on the power of the wave at any point in space. Because of the myriad of variables affecting the wave, it is impossible to know with certainty the exact signal strength the wave will have at a particular point in space. Statistics plays a big role in understanding and defining the "average" behavior of the wave in an environment.

The frequencies we are interested in are those above 700 MHz, since this is where adequate amounts of spectrum have been assigned to support high bandwidth systems. The upper limit to spectrum useful for a non-line of sight communication path is around 6 GHz. Above this frequency, radio waves behave more like light, and no longer refract around objects in the path. Frequencies in the 10 to 70 GHz bands are very useful for building point-to-point communication links that can be used to connect communications sites back to a hub location for traffic aggregation. They are also useful for extending high bandwidth connections from one location to another.

Radio waves, like light waves, get weaker with distance. The attenuation associated with distance in an unobstructed path is called *Free Space Loss (FSL)*. FSL is mathematically calculable by the formula 20Log_{10}(Frequency in MHz) + 20Log_{10} (Distance in Miles) + 36.6, because it is the result of the spreading of the wave as it propagates away from its source. The attenuation is also frequency dependent. The higher the frequency, the more attenuation will occur over a given distance. As you see in Figure 3-1, the loss change follows a 6 dB per octave (a doubling

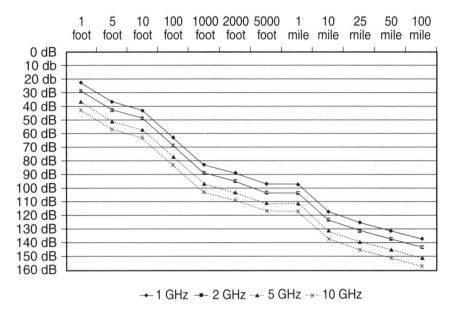

Figure 3-1: Chart showing loss vs. distance at 1, 2, 5 and 10 GHz

of frequency or distance) or 20 dB per decade (a ten fold change in frequency or distance) slope.

Since Free Space Path Loss can also be described as a 20 dB per decade loss, it means if the signal is –104 dBm 1 mile from the transmitter, it will be –124 dBm 10 miles away, and –144 dBm 100 miles away. A useful gauge for real-world application is: A 6 dB power change will double or halve path distance. The reason that radio waves attenuate in this manner is because, like light, the wave front continues to expand spherically, thus, as shown in Figure 3-2, the wavefront energy is spread over an ever widening area, thus reducing its density at any single point in space.

When discussing free space loss, more than just optical line of sight must be considered. It's obvious that an unobstructed path contains no objects that block the optical line of sight. However the real world of radio is a bit more complex than that. The unobstructed path needed for Free Space Loss requires the path to be both optically clear and have a clear Fresnel zone surrounding it.

The area between the transmitter and receiver can be defined by a series of concentric ellipsoids that correspond to the ever-widening field of the radio wave. The term Fresnel zone defines the shape of these ellipses as a circular zone with

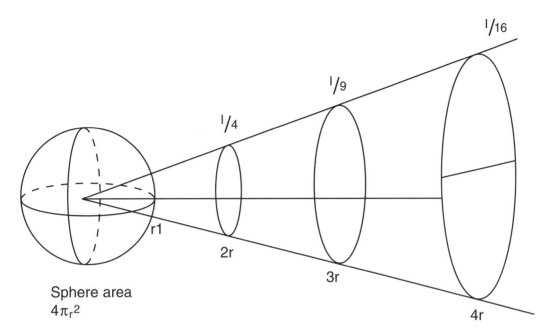

Figure 3-2: Expanding wavefront and area power density

a radius such that the distance from a point on this circle to the receiving point is some multiple of a half wavelength longer than the direct path.

$$\text{Fresnel zone} = 43.3 * \text{SqrRoot} (d_{Mi} / (\text{Freq}_{GHz} * 4))$$

Why is a clear Fresnel zone important? Because radio waves reflect off objects much like light reflects off a mirror. Figure 3-3 shows the effect of having an

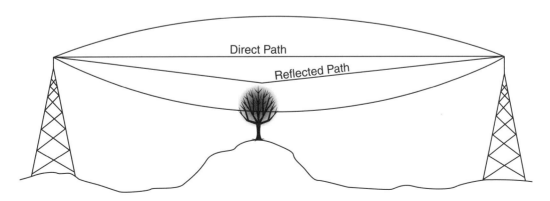

Figure 3-3: Fresnel zone

object outside the optical line of sight, and within the Fresnel zone: reflections are generated. Defining the Fresnel zone as a point with a distance a multiple of a half wave longer than the direct path is done with good reason. An object in the Fresnel zone gives rise to a reflection. That reflection also propagates toward the receiver, only it arrives later than the direct wave and out of phase with the direct wave by half wavelength intervals. Since radio signals are sine waves, being a half wavelength out of phase causes wave cancellation.

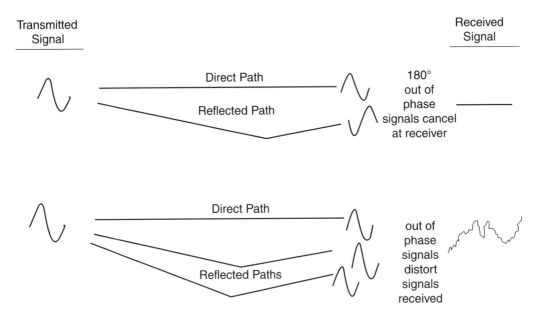

Figure 3-4: Sine wave and phase diagram illustrating wave distortion

Also, since the direct wave and reflected waves are traveling over different distances, they arrive at the receiver displaced in time and phase. Since the direct and reflected waves are carrying the same information, the temporal and phase displacement results in distortion of the received wave. These effects are illustrated in Figure 3-4.

These destructive effects are known as multipath interference. The result of multipath interference is that the arriving wave is no longer pristine. It has been affected by random cancellations that reduce its signal strength, and the distortions introduced by the temporal shift in the arriving waves have led to more uncertainty about the actual phase and amplitude states contained in the arriving wave.

The net result is a loss of signal integrity that leads to an increasing error rate at the receiver. A single Fresnel zone incursion can degrade the signal at the receiver by 10 dB or more. That's enough to reduce your system's coverage distance by ten-fold!

These reflections also introduce signal fading into the radio link. Because of movement of the objects causing the reflections (like leaves in the wind) the radio path exhibits fluctuations known as fading, which results in random variations in amplitude and frequency response. Fading effects are illustrated in Figure 3-5, and can manifest itself in two forms: log normal fading and frequency selective fading. If the radio channel has a constant gain and a linear phase response over a bandwidth larger than the bandwidth of the transmitted signal, it is exhibiting log normal or *flat fading*. Log normal fading can be seen as a variation in amplitude of the entire signal. Under these fading conditions, the received signal has amplitude fluctuations due to the variations in the channel gain over time caused by multipath. In flat fade conditions, the spectral characteristics of the transmitted signal remain intact at the receiver, only the amplitude varies. In other words, the entire channel fades as a constant.

On the other hand, if the radio channel has a constant gain and linear phase response over a bandwidth smaller than that of the transmitted signal, the

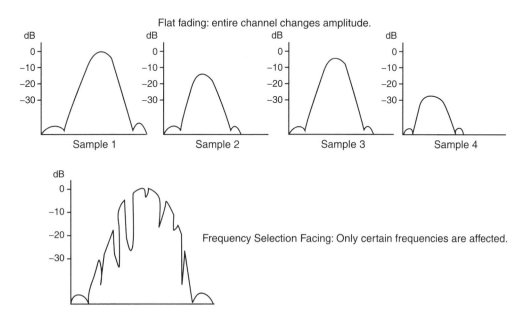

Figure 3-5: Fading diagram showing flat and frequency selective fading

transmitted signal is undergoing *frequency selective fading*. In this case, the received signal is distorted and dispersed by the multiple arrivals of the main and reflected signals. The result is that the channel no longer fades in its entirety. Only certain discrete frequencies are affected. The affected frequencies are the direct result of half wavelength multiple temporal and phase displacement of the arriving signals. Only those discrete frequencies within the channel that happen to have an exact ½ wavelength multiple relationship with each other are affected. The result of frequency selective fading is that the received signal has "holes" across the channel. Frequency selective fading can cause 30 dB or more of attenuation at certain discrete frequencies in the channel.

It is interesting to note that fading effects are very localized. Fading effects differ over distances of about half a wavelength. In many cases moving the receiver or transmitter by a half wavelength or more will cause a completely different signal to appear at the receiver.

You may have noticed this behavior on your cell phone. There are many areas where the signal may be good in one location and poor just a few inches away. Turning your head can sometimes mean the difference between a conversation and noise.

Multipath and its attendant fading cannot be eliminated, but its effects on signal integrity can be lessened by increasing link power or receive sensitivity. This can be accomplished by: additional antenna gain, additional transmitter power, simpler modulation (with reduced throughput), or moving the remote station closer to the base station.

Since the FCC sets maximum power limits on communication channels, there will be limits associated with maximum path lengths supportable by a given technology and modulation. This is especially true of those associated with part 15 use, where the power outputs are quite low in order to allow operation by a myriad of devices without requiring usage coordination. If additional power or antenna gain is not available or useable, then reducing modulation complexity and its attendant reduction in throughput may be the only way to make the link work.

Fresnel zone clearance is a term normally associated with point-to-point microwave links. By designing the link to provide a clear Fresnel zone, free space loss can be assumed as the path loss attenuation factor for the link. If a clear Fresnel zone cannot be achieved on a point-to-point link, then larger antennas or more transmit power are necessary to assure that the link can overcome the effects of multipath and provide the quality of service it was designed for.

In a system with a base station serving multiple end user clients, like a wireless LAN and mobile/portable communication systems, the remote ends (or the users) are located within a field of three-dimensional objects. In such a case, all those objects become Fresnel zone incursions in addition to being potential obstacles to optical line of sight. This leads to an environment where multipath can come from a large number of sources, as shown in Figure 3-6. In the simplest case, a lap-top computer on a desk in an office, there are multiple reflective surfaces. Walls, file cabinets, people, windows, ceilings, floors, even the desk and laptop itself can cause signal reflections that affect the integrity of the signal as it's received. Multipath interference also exists in an outside environment, where buildings, vehicles, roads, trees, and people provide the reflective surfaces.

Multipath is a fact of life in mobile or portable system design. Because the re-mote end is always located among a mix of obstacles and reflective objects, and generally the elevation of the remote end antenna is lower than surrounding obstructions, the effects of multipath and path blockage must be considered when designing the system.

So far we've just seen what effect peripheral objects have on the signal path. What happens when an object obstructs the path completely? In addition to some of the signal being reflected, the signal going directly through it is attenuated and some of the signal going around the objects edges is refracted. In some cases where the direct path is blocked by an obstacle, the only available signal at the receiver is the result of a reflection or refraction.

The amount of signal that is attenuated by an object depends on its size and the material of which it is constructed. Common building materials and their attenua-tive properties are listed below:

Plasterboard	3 dB to 5 dB
Glass wall with metal frame	6 dB
Cinderblock wall	4 dB to 6 dB
Window	3 dB
Metal door	6 dB to 10 dB
Structural concrete wall	6 dB to 15 dB

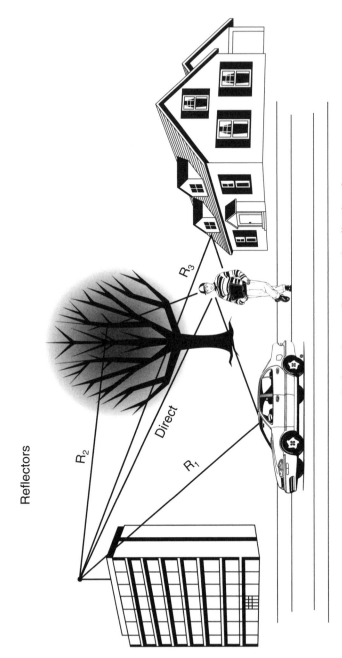

Figure 3-6a: Typical outdoor reflectors of radio signals

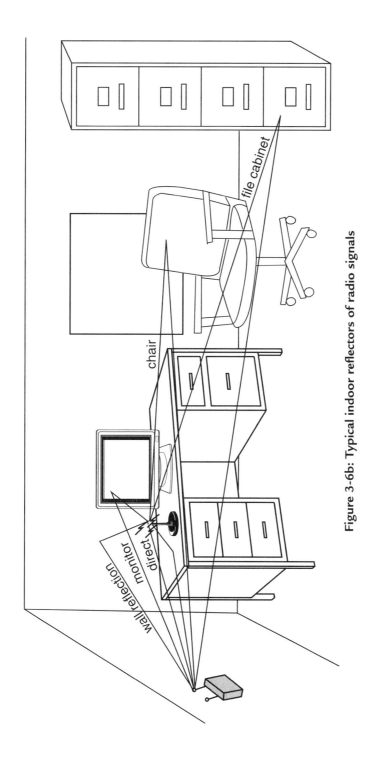

Figure 3-6b: Typical indoor reflectors of radio signals

These numbers relate to each instance of the signal penetrating this construction material, therefore in a typical small office environment the signal may be penetrating several plasterboard walls resulting in an additional 9 to 15 dB of attenuation. Metal doors or structural walls could also be obstructing the signal, causing even larger amounts of attenuation.

The effect of these random obstacles on the propagation path is known as shadowing. Shadowing is normally of great concern when designing portable or mobile systems because shadowing can lead to localized signal strength changes on the order of 10 to 30 dB, depending on the environment. Shadowing occurs when the remote end is moved a sufficient distance so as to cause gross variations in the overall path between the transmitter and receiver. The effect is called *shadowing* because the remote end has moved into an area that results in it being in the "shadow" of surrounding objects. Due to the effect of multipath, a moving receiver can experience several fades in a very short duration, or in a more serious case, the remote end may stop at a location where the signal is in deep fade. In such a situation a slight movement of the remote end antenna location may put the receive antenna into a better location. Another way to accomplish the same thing automatically is through the use of diversity receive antennas. Some 802.11 access points and client cards have this feature. If a single-band access point has two antennas on it, it most likely includes diversity. Many client cards also have diversity antennas, but because they are hidden within the shell of the card, they are impossible to see. Diversity is commonly deployed in portable and mobile systems in order to improve the system's ability to deal with fading and multipath.

Since fading is a localized phenomenon, diversity works by allowing the receiver to listen on two independent, spatially separated antennas. If the antennas have enough physical separation, each will see an independent signal that has completely different fading characteristics. The receiver is then able to choose the better of the two signals. In order to have completely uncorrelated signals at each antenna the antennas need to have at least 10 wavelengths separation. In this ideal situation, diversity can improve the receive signal strength by 3 to 6 dB. Diversity will work with significantly less antenna separation; even a ½ wavelength separation will provide some signal strength improvement.

Path Loss and System Coverage

Now let's take a look at how all these variables can affect the coverage offered by a radio system, and begin to look at the tools you will need in order to analyze, design and manage a system.

The first thing you need to do is determine how many dB of radio path loss the equipment can tolerate and still function. This is easily calculated once you have a few basic bits of information about the equipment you plan to use.

You will need to know the transmit power, the receive sensitivity, the antenna gain, cable and connector losses, and diversity gain of each station (i.e., base station and client equipment). These factors are used to calculate a Link Budget for the equipment by using the formula:

$$L \ [dB] = Ptx \ [dBm] + Gtx \ [dBi] - Prx \ [dBm] + Grx \ [dBi] + Gdv \ [dBi] - M \ [dB]$$

Where *L* is the link budget in dB, *Ptx* is transmit power, *Prx* is receiver sensitivity, *M* is fading margin, *Gtx* and *Grx* are antenna system gains on the transmit side and receive side respectively, and *Gdv* is diversity gain. Antenna system gain includes both the gain associated with the antenna and the loss associated with the coax cable and connectors that feed it. Feedline and connector loss is one of those easily overlooked sources of loss that can make the difference between a reliable system and one that may not work at all. Above 1 GHz, the impedance imbalance associated with each coax connector can easily add 0.5 to 1 dB of loss to the system. If the connectors are of poor quality or are improperly assembled, the losses per connector can rise to over 3 dB! Feedline loss is also significant at frequencies above 1 GHz. Even really good cable like LMR-400 coax exhibits 6.8 dB of loss per hundred feet at 2.4 GHz and 10.6 dB at 5.6 GHz. Keep the coax run as short as possible, and make sure to use high-quality, properly-assembled connectors.

It is important to look at both the base to client (downlink) and client to base (uplink) communication paths. Often, the base station will have a more sensitive receiver, a more powerful transmitter, and diversity, while the client end will have lower performance due to size and power constraints. This can lead to an equipment design that is "unbalanced," meaning it can "talk" farther in one direction than the other. Ultimately, the useable path will be defined by the link direction having the lowest performance. The link budget calculator spreadsheet included on the CD-ROM is an expansion of the previous equation and is useful for analyzing the performance of both uplink and downlink paths, and recommending power settings to achieve path balance.

The link budget is a critical part of the design, because it tells you how much path loss can be overcome by the system. This will directly relate to the coverage available from the system in any environment. This one number is the key to determining how well the equipment will work in a given environment.

For example, let's design a point-to-point communications link with off the shelf 802.11b hardware. Common operating characteristics are: transmit power +17 dBm, Receive Sensitivity at 11 Mbps –84 dBm, 3 dB of coax losses in the cable that connects the equipment to the antenna. Since this is a point-to-point facility, a single 18 dB directional antenna will be used at each end, thus no diversity gain is available. A 10 dB fade margin will be used to account for environmental factors like weather. Since the equipment at both ends of the link are identical, only one path loss calculation needs to be performed. If there is different equipment at each end, then a calculation needs to be run for both directions. Select the "worst" performing direction as basis for further analysis.

Using these numbers in the equation yields the following:

$$L = 17\ dBm + 15\ dBi - (-84\ dBm) + 15\ dBi + 0\ dB - 10dB = 121\ dB$$

This means that the radio link can overcome 121 dB of attenuation and still remain functional.

Assuming that this point-to-point link has a clear Fresnel zone allows us to use the free space loss formulae $20Log_{10}$(Frequency in MHz) + $20Log_{10}$ (Distance in Miles) + 36.6. Because this is 802.11b equipment in this example, the operating frequency of 2.4 GHz is used in the formula.

The result of the calculation shows that this link can span a distance of 6.7 miles reliably. Of course to achieve this distance you'd need to do whatever was necessary to make sure the link was operating with a clear Fresnel zone, because if there is ANY clutter in the path that encroaches on the Fresnel zone, according to Figure 3-7, you can add an additional 16 dB of loss to the path. So using our handy references, a 10 dB change reduced the propagation distance by an order of magnitude, and a 6 dB loss halves the distance, so the coverage drops to 0.335 miles, all due to an object in the path that didn't even interfere with optical line of sight!

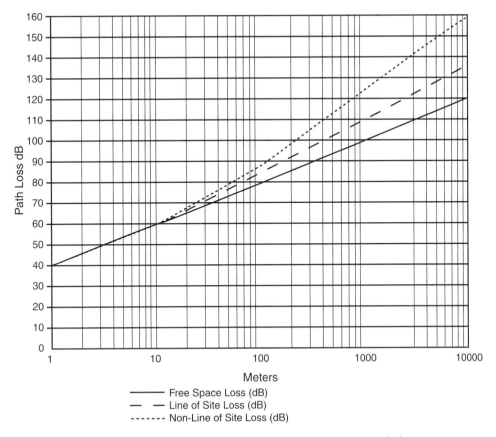

Free Space Loss (dB)
— — Line of Site Loss (dB)
------- Non-Line of Site Loss (dB)

Figure 3-7: Free space vs. line of sight vs. non-line of sight graph for 2.4 GHz

You can see in Figure 3-7 that free space loss offers a huge range advantage. Optical line of sight with an encroached Fresnel zone offers considerably shorter paths. The path gets shorter still if an object actually blocks the optical line of sight in the path, as is the case with the non-line of sight curve in the table.

This example was useful for a point-to-point link, but what about a base station to client link? This is the type of system that you will normally encounter when designing a wireless LAN or a Wireless ISP facility. In these systems, the user will be located in an environment that is filled with multipath and shadowing effects. Since you can no longer consider only one remote end to the facility, predicting the attenuation over the aggregate coverage area becomes important. This is much easier said than done. The number of interactions in the environment becomes almost infinite, and therefore impossible to predict with complete accuracy. Still, from a planning standpoint, using an average loss calculation is useful for estimating coverage and the amount of equipment necessary to cover the given area.

Using the path loss formulae and the path loss attenuation estimators will allow you to estimate the expected coverage of any equipment you might be considering. This is valuable from several standpoints: it allows you to compare the performance of competing hardware, it allows you to estimate how many base station locations will be necessary to cover the intended area, and it gives you a benchmark that is based upon physics and your actual operating environment instead of equipment marketing hype. Over my career, I've seen too many instances where equipment was represented based upon an ideal operating environment. Unfortunately, this environment is rarely available in the real world.

A word of warning: the path distance estimations are just that: estimations. The actual coverage may vary wildly from the estimates if you did not accurately calculate the effect of obstacles in a non line of sight path. Even if you did do a good job of determining the attenuators, there will still be location specific variations in coverage that are based upon the real interaction of objects in the environment. Shadowing and fading effects will compound the loss in certain areas but not in others. The result is that the actual covered area will be irregularly shaped, as seen in Figure 3-8.

The physical environment encountered by a radio device operating in non-Free Space Loss conditions is anything but benign, and anything but predictable. This is especially true in a system serving multiple portable or mobile devices. In the real world, actual coverage will vary from the estimates generated using simple path loss equations. This is because the path changes at every point within the covered area. To predict the propagation of a signal accurately, you need to know the gain and loss of each part of the system, including the loss through the propagation medium. While equipment specifications are easy enough to find, determining the path loss can be tricky.

The factors affecting propagation losses in an indoor environment are multipath effects, local clutter and the construction materials used in the environment. In an outdoor environment, terrain and morphology (land use) have the greatest effect on propagation losses.

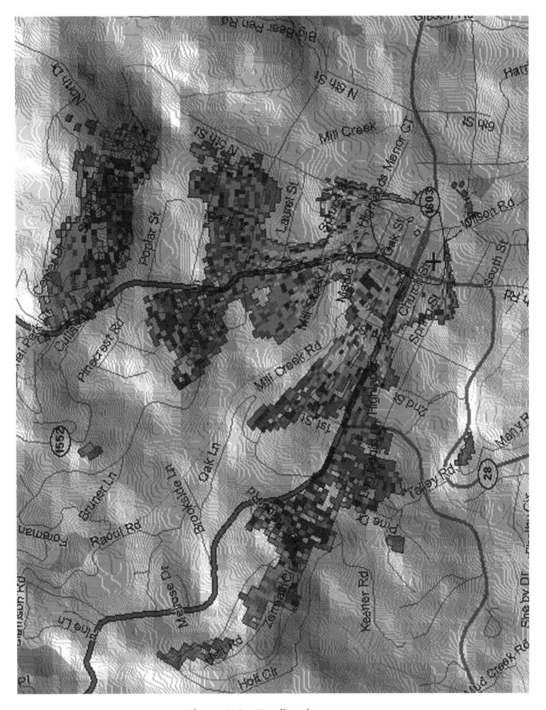

Figure 3-8a: Predicted coverage

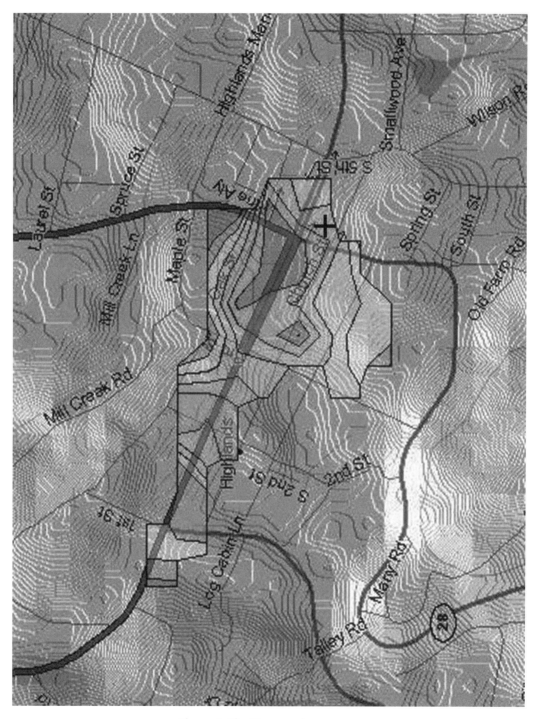

Figure 3-8b: Measured coverage

The uneven propagation losses across the covered area affect not only desired coverage, but undesired coverage as well. How can coverage be undesired? If a "foreign" signal is on the same frequency as the desired one, it can cause an increase in noise or cause interference. This causes a reduction in system performance because of the deleterious effect of having an undesired interfering signal in the coverage area.

Frequency Reuse

Radio spectrum is a scarce resource, so there is never enough available to allow every radio facility to have it's own unique channel. Even radio and TV broadcasters "reuse" the limited pool of assigned frequencies on a city by city basis.

The concept of frequency reuse is simply the use of the same frequency in multiple locations where there is sufficient physical separation so as to allow interference-free operation by all users of the frequency. In TV broadcast you can see this effect in the assignment of channels along the Eastern Seaboard. New York City broadcasters are allocated channels 2, 4, 7, 9, and 13. Philadelphia, 90 miles away is allocated channels 3, 6, 10, and 12. Washington DC, 170 miles from New York is sufficiently distant that channels 2, 4, 7, and 9 can be used again because the interference contour generated by the either station never overlaps with the desired coverage area of either station.

The same concept is used when deploying a fixed or mobile data network, although the co-channel separation distances are much shorter. When planning a system that has more base stations than channels, a reuse plan must be implemented. If you are designing point-to-point (PTP) links, antenna aperture can be used to control the coverage area of a site, and make channel reuse possible at very close distances.

In a PTP link, you know exactly where both ends of the communication channel are located, so you can focus your RF energy at one unique point in space. You do this by selecting very high gain directional antennas. By their nature, these antennas have very narrow apertures, sometimes as narrow as a few degrees. As you see in Figure 3-9, these antennas have a very narrow beamwidth and outside the main beam the signal strength falls rapidly. This provides two benefits: first, the receivers do not see all the noise and interference in the environment, only what exists within

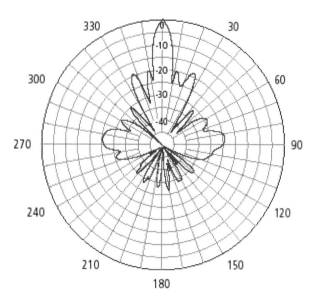

Figure 3-9: Antenna pattern of a microwave parabolic dish antenna

the narrow aperture of the antenna. Second, the transmit energy is contained in a very small area, so it does not generate interference over a large area.

With the proper antennas, power settings, and angular separation of stations, one can reuse the same channel to serve several links originating on the same rooftop. This concept is seen in Figure 3-10.

The situation becomes more complex when reuse is required in a system that provides connectivity to various fixed, portable and mobile users interspersed around a coverage area. In this situation the antenna needs a wide field of view in order to serve multiple users. The propagation over this wider environment will be less uniform, leading to compromises to such factors as power, base station location, and coverage being made. The key in this environment is to get as much coverage in the desired area as possible while limiting the amount of RF energy extending beyond the desired area.

In a purely theoretical world composed of flat space, each base station would have a uniform circular coverage pattern. The size of the circle would be based upon the height of the antenna and the maximum path loss of the system. Adjacent

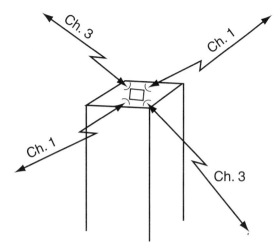

**Figure 3-10: Illustration of reuse on
a common rooftop**

sites would be placed so as to have the minimum overlap necessary to assure that all areas were covered without gaps or holes in the coverage. These overlapping circular coverage areas can be viewed as hexagons, with each hexagon side defining the location where signal from adjacent base stations are equal. Although such conditions don't exist in the real world, the exercise is still useful for system planning, because it allows you to visualize coverage of individual base stations, and identify the best fit location for all the neighboring base stations. Figure 3-11 shows the progression from circles to hexagons and shows the real coverage achieved from a site. As you can see, the real-world coverage is neither circular nor hexagonal. Instead it is amorphous, its shape being determined by the environment through which the signal propagates.

For now let's focus on assigning channels to the individual sites, and creating a reuse pattern. The primary purpose of the reuse pattern is to divide and assign the available frequencies in a regular repeatable pattern that separates the co-channel and adjacent channel users sufficiently as to control interference. Note that I said control, not eliminate, interference. If reuse distances were so great as to eliminate all possibilities of interference, then spectral efficiency would suffer. Assuming you have enough channels, the key is to find a spacing that is adequate enough to provide sufficient C/I or Eb/No (energy per bit/noise in channel) protection to the majority of the desired coverage area.

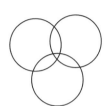

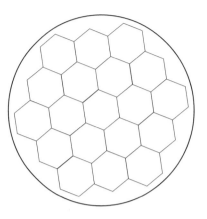

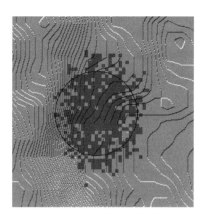

Figure 3-11a: Simple circles can define a covered area for planning purposes.

Figure 3-11b: Overlapping circles are more easily gridded as hexagons. This is useful for larger networks and coverage areas.

Figure 3-11c: But real-world coverage is amorphous and will not match simple geometric planning guidelines.

Reuse patterns of 1,3,4,7,9,12,and 21 are commonly used in commercial telecommunications networks like cellular phone systems. The reuse pattern can be used with omnidirectional antennas, or, if more protection and capacity is needed, directional antennas (and a larger channel set) may be used. The reuse pattern is selected based upon the C/I needs of the technology in use on the network. As Figure 3-12 shows, these reuse patterns are nothing more than a regular, repeatable distribution of channels to each base station. By using such a pattern you can grid out your system in such a way as to generally provide separation between co-channel base stations that is sufficient to assure the interference levels will not negatively impact the users of the channel. Of course, as you saw in Figure 3-11c, the actual coverage is never perfectly fit within the simple circular or hexagonal grid, so opportunities for interference between sites still exist, even in the best planned system. This opportunity for interference could be eliminated by selecting a reuse plan with more spacing between co-channel sites, however this would reduce the capacity of the network, lead to inefficient use of spectrum, and increase the cost of the network.

The amount of interference generated in the reuse area is higher for smaller reuse patterns. As you can see in Figure 3-12, a reuse pattern of 1 offers little C/I

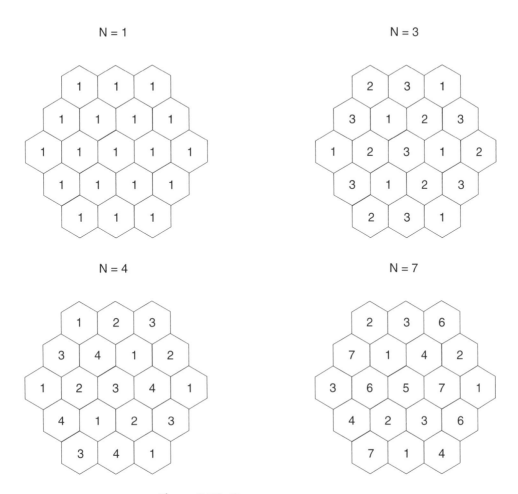

Figure 3-12: Common reuse patterns

protection: near the edge of coverage, powers are nominally the same from both base stations. This pattern is used in CDMA systems that utilize dynamic power control (DPC) to minimize interference. Even with DPC, no single base station can reach its full capacity or coverage potential when it is surrounded with co-channel reuse sites.

Often, an equipment manufacturer will state the minimum C/I, Eb/No or SNR requirements of their equipment, and may even recommend a reuse pattern to use with their equipment. Optimized reuse patterns are commonly found in commercial radio networks. Reuse can be optimized to minimize interference because the spectrum is licensed to and managed by a carrier.

N reuse Pattern 1	Omni C/I 0	Omni Channels Required 1	120 degree C/I 2	Sector Channels Required 3
3	7	3	11	9
4	9	4	14	12
7	14	7	19	21
9	16	9	21	27
12	19	12	23	36
21	23	21	28	63

Approximate achievable C/I and channel requirements for various reuse patterns

Using the unlicensed bands leads to the inability to manage a reuse plan because you are not the only user of the band. Even if you could find sufficient channels to create an optimized reuse plan, the fact that others are using the same spectrum means that you cannot optimize the reuse plan to suit both your internal frequency requirements and those of other users around you. For this reason, unlicensed frequencies will have great difficulty providing ubiquitous high quality area coverage.

Current public standards for unlicensed spectrum, like 802.11b were initially designed for indoor LAN extension applications. There was more consideration given to efficiency than was given to supporting close spaced reuse. The result of this is that the 2.4 GHz band supports only three nonoverlapping 802.11b channels.

This means that, like it or not, a reuse pattern of 3 is the best you can get using 802.11b equipment. This will cause some limitations vis-à-vis interference free operation. This will be especially true when deploying networks in multistory buildings.

The reuse examples we've looked at so far have only considered reuse in two-dimensional space. When a third dimension is added, the world gets more complex. In a multistory building, each floor can use the N = 3 reuse pattern. This may work OK on a single floor, but what happens on floors above and below? Since, desired or not, there will be propagation between the floors, this leads to another source of interference that needs to be considered in the reuse pattern used on each floor.

Think of each base station in this multifloor building as having an ovoid coverage pattern. It won't be circular because of the added attenuation of the floor and ceiling. The largest coverage area will be on the intended floor, but the top and bottom of the ovoid will extend to the floors above and below the covered floor. Let's say that each floor requires four base stations to provide coverage. That means in an 802.11b network that one channel will need to be reused on each floor. In other words one of the three available channels will provide coverage to 50% of the floorspace on the target floor. Now that same reused channel has the potential of causing interference to 50% of the floorspace above and below it. This severely limits the selection of interference free reuse channels on these floors because you have to manage not only the interference on the floor, you also have to coordinate and manage interference from the adjacent floors!

Obviously, if there is sufficient propagation through the floors and ceilings, there will be unacceptable interference from one floor to another. Channel usage considerations in such an environment will require an understanding of the propagation characteristics, user throughput requirements and traffic characteristics. These concepts will be further discussed in Chapter 5.

CHAPTER 4

Propagation Modeling and Measuring

- Predictive Modeling Tools
- Spreadsheet Models
- Terrain-Based Models
- Effectively Using a Propagation Program
- Using a Predictive Model
- The Comprehensive Site Survey Process
- Survey Activity Outline
- Identification of Requirements
- Identification of Equipment Requirements
- The Physical Site Survey
- Determination of Antenna Locations
- RF Site Survey Tools
- The Site Survey Checklist
- The RF Survey
- Data Analysis

Propagation Modeling and Measuring

When designing a system or network, it is helpful (if not imperative) to know the coverage area provided by each site. This is important for two reasons: first, you want to assure that the users in the desired coverage area are served with a high quality signal, and second, you need to know how each transmitter adds to the interference levels in surrounding areas.

In order to evaluate the coverage that will be provided by the selected hardware, either propagation prediction models or physical surveys can be used. Propagation modeling is accomplished with a software tool, while physical surveys are accomplished by temporarily installing hardware then measuring the resulting coverage. If you are building few transmitter locations, or are only constructing systems inside buildings, the site survey may be the quickest and is certainly the most accurate method. If on the other hand, you are planning multiple outdoor sites in various areas, the time and expense associated with acquiring and learning a software-based predictive model may prove valuable.

Predictive Modeling Tools

In the early 1980s, the first large scale cellular telephone networks began to be planned and constructed. In order to support these projects it was necessary to develop tools that would allow reasonably accurate prediction of RF propagation. Without such tools the only way to assess the coverage of a site was to perform a lengthy, complex and costly "drive test" on each site alternative. This involved erecting an antenna at the appropriate height (often 100 or more feet in the air), connecting a transmitter operating at the appropriate power, and then driving around the desired coverage area with a special receiver capable of recording signal strength and location. Obviously, this was a massive amount of effort when contemplated for thousands or tens of thousands of locations.

Luckily, during the 70s a significant body of work defining the statistical properties of RF propagation was accomplished worldwide. This work led to the development of a series of algorithms that described the mean behavior of RF over a varying environment, terrain, and morphology. In their simplest forms, these algorithms described a series of curves that identified the propagation loss per decade over the various environments.

As cellular systems matured, the complexity of cellular systems increased and additional bands at higher frequencies became available. This led to additional refinement of the algorithms used to predict propagation. These propagation models have names like TIREM, HATA, Longley-Rice, and Walfish-Ikegami. All are differing attempts to characterize propagation in different environments accurately.

It is important to realize that these propagation models are, at best, estimations of real-world propagation. The models are based upon statistical behavior, and the data that represents the terrain and morphology that the signal propagation is evaluated over is rather coarse and incomplete. Still, the models are useful for analyzing coverage and interference if you understand, accept, and design around their limitations.

Spreadsheet Models

In their simplest form, these algorithms can be implemented in a spreadsheet that is used for calculating the average coverage area of a communication site for the purpose of estimating the number of sites necessary to provide service in the designated area. Though they provide a more realistic estimation of coverage than simple free space loss calculations, they still do nothing more than allow you to draw a circle describing the range of a system. Figure 4-1 shows the different ranges achievable based on varying the path loss slope. Obviously, it is important to understand the average loss characteristics of the environment before selecting a loss slope. Selecting an inappropriate slope will lead to significant over or under estimations of the actual propagation in the area.

This simplistic approach can be useful as a financial planning tool because it allows the approximate system costs to be known, but it is still not useful for designing a network that would provide known coverage in a known area, as well as predict interference levels outside the desired coverage area. For this, more complex models are necessary.

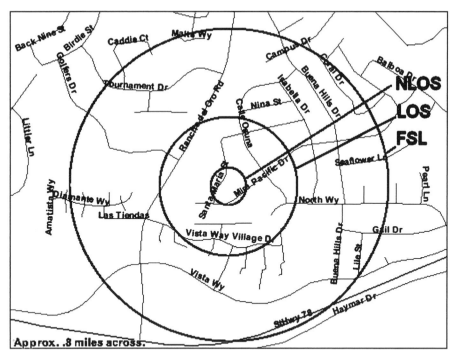

Figure 4-1: Map with different circles describing coverage based on various loss slopes

Terrain-Based Models

These models needed to take into account the actual terrain and morphology (land use) within the coverage area, and calculate coverage based upon those characteristics. These models begin to represent the actual behavior of the RF signal based on what obstacles it encounters while propagating outward from the transmitter. The same algorithms discussed above form the basis for these more complex models. The difference is that the models are applied over known terrain and (maybe) morphology. This is accomplished by using digital terrain data. Terrain data is available from several sources, including the USGS. Such publicly available data has a resolution of 1 km, 100 meters, 30 meters, and 10 meters. This means that the data is averaged into a block of the size shown. 100 Meter data averages all the terrain in a 100 × 100 meter square, and represents it as a single elevation. 10 meter data averages the terrain contained in a 10 × 10 meter square. This averaging does lead to some inaccuracy in coverage prediction, so it is important to acquire the highest resolution data available. Morphological data is harder to find. It is normally custom digitized from high altitude stereometric photographs. Morphological data is also somewhat time limited in its utility, because trees grow and new buildings are constructed.

Modeling is accomplished by placing the simulated radio base station in a modeled environment representative of the actual area to be covered so signal strength could be predicted. This is done by looking out across the digital landscape represented by the terrain and morphological data, and calculating the mean signal strength based upon the environment present in that single slice. Each model predicts the mean loss using different parameters, but the results are the same: a plot identifying the expected propagation overlaid on a terrain map, or road map, or both.

In order to accomplish this, large scale computing power is necessary. The first of these models were run on mainframe computers, and as processing power increased, minicomputers. By 1984, programs were developed that could run on a PC. Today many programs exist and are available for purchase or license from various sources. All of these programs can run effectively on a modern desktop or laptop computer.

Effectively Using a Propagation Analysis Program

Acquiring a software package and operating it is not the hard part, effectively using it is. There are a number of issues that you must consider in order to assure the propagation predictions reflect the real world. You must make sure you have accurate terrain information and morphological information. In addition, you should do field measurements from a number of sites in various settings and compare the measured results to the predicted results using the technology and band you are implementing. This comparison will show whether the model over or under predicts, and allow you to "tweak" variables in the model in order to make the predictions line up with reality. By doing this you gain confidence in the model, and eventually you will be able to rely on propagation modeling to predict the behavior of new sites without always resorting to field tests and site surveys.

Propagation prediction software is a tool most effectively used by engineers with some experience in RF propagation. Without a certain level of experience, the tool provides little value. For example, propagation prediction software is only as accurate as the underlying information in the database it uses for predicting coverage. It is important to acquire the highest resolution terrain data available. In addition to terrain, the land has usage, or morphological features, such as buildings, roads, and trees. If these factors are not considered in the database, then accuracy suffers. Take, for example, New York City. Because the terrain is relatively flat, if you were to use terrain alone to predict coverage from a site with antennas at 100 feet elevation, you would predict large coverage area from any site in Manhattan. The actual propagation would be far less due to the high density of buildings in the area.

The coverage shown in Figure 4-2 shows the difference between predictions based on terrain only, and terrain plus average building clutter. While the propagation prediction is more accurate with average building clutter data added to the terrain, it is still inaccurate. The actual coverage from the site in question would look more like a "+" sign. The site is located at a street intersection, so there is no blockage along the street in the north-south and east-west orientations. Therefore the signals will propagate significantly further in those directions than in any other orientation. The predictive model did not identify this because the average building clutter data and terrain data did not have enough resolution to identify roads vs. buildings. Manhattan is an area where the use of data acquired from stereo photography may be relevant. Because the digitized image contains information about the actual location and orientation of streets and buildings, as well as the actual height of buildings and the level of the streets between them, it provides a much more accurate representation of the real world. The propagation software would be able to "see" the actual environment as a series of deep narrow canyons (streets) amongst tall dense canyon walls (buildings). Using this type of data instead of average terrain and morphology will allow significantly more accurate propagation prediction. The downside of this approach is cost. The database generated from the stereo photograpy is quite expensive to obtain.

While Manhattan is an extreme example, the same inaccuracies propagated by the use of average terrain and morphology will exist in any area where there are natural or manmade objects on the ground. These objects affect propagation by both blocking the radio path and by providing reflective surfaces that cause multipath. Both these effects lead to signal attenuation in the real world that, depending on the accuracy of the terrain and morphological databases, may not be appropriately considered by the propagation model.

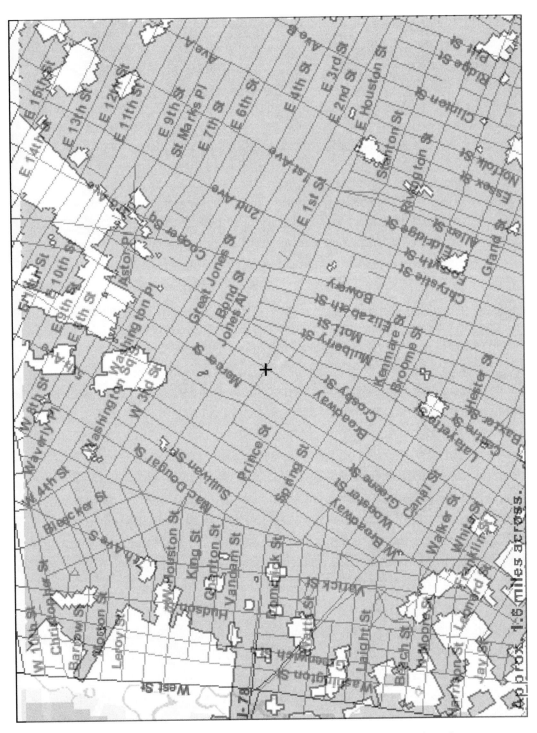

Figure 4-2a: Predicted Manhattan coverage based upon terrain only

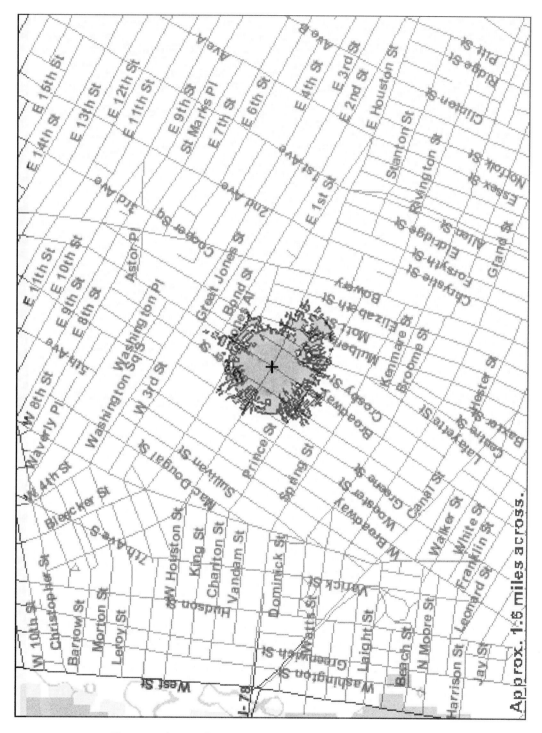

Figure 4-2b: Manhattan coverage based upon morphology

This is why field surveys are necessary. The survey provides actual coverage information that can be compared to the predictive model. By comparing the two results you can begin to understand the average propagation of radio signals in different environments. In fact, this is how the predictive models were initially created. Millions of measurements were taken in different environments, and the results were graphed as signal strength vs. distance from the transmitter. Statistical analysis was done on the results, and a determination of the average loss over distance and impact of diffractive and reflective objects was determined. These statistical results were then represented as a set of algorithms that are used in the predictive models to characterize propagation across a given environment. Of course, you will not need millions of data points. We're not trying to create a new algorithm, just verify the accuracy of the one we've chosen.

By using a combination of predicted propagation, field surveys, plus intuition and logic, you can estimate the coverage available from a site. Let's take the Manhattan example. The prediction based on terrain alone is useless for determining general coverage. It is, however, representative of the coverage achieved on the streets in direct view of the site, thus the "+" shape defines the extended coverage area associated with the immediate cross streets. The buildings surrounding the site will attenuate the signal by over 25 dB each, so you can assume that there will be some in building coverage in the buildings directly adjacent to the site, but the intervening buildings will attenuate the signal so much that the signal will not appear in buildings a block away. Given these assumptions, you could estimate the coverage to be similar to Figure 4-3.

The astute reader may have noticed that I have shown coverage expectations for the above example without representing a number of key factors. What frequency was used? What antennas? How much transmit power? What receive sensitivity? What cable loss?

In addition to terrain and morphology, these factors need to be known in order to effectively use any propagation model. As discussed in Chapter 4, these are the critical factors for determining maximum path loss, and are among the key variables used by the predictive modeling software to calculate coverage.

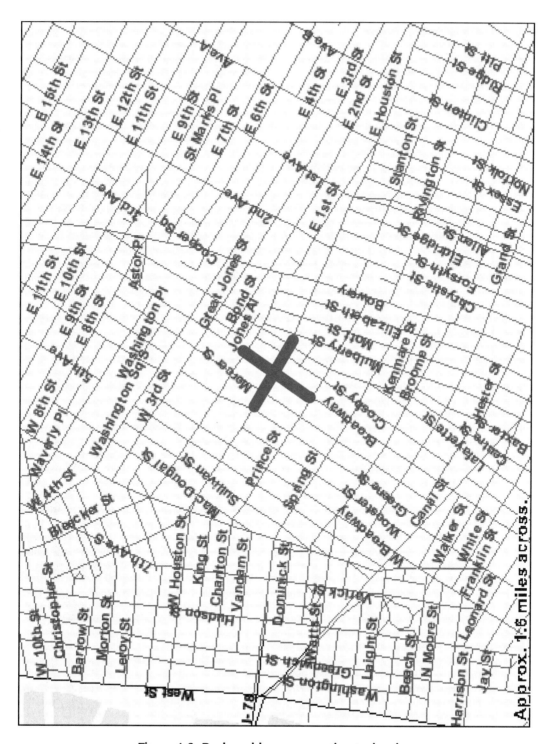

Figure 4-3: Real-world coverage estimate drawing

Using a Predictive Model

Let's take a look at some actual propagation modeling software, and load it up with real-world data and system variables. The software that generated the propagation plots in this book is a program called *Radio Mobile Deluxe*, written by Roger Coudé and made available as freeware. A copy of this program has been included on the CD-ROM accompanying this book. Roger is constantly updating the program and adding new improvements, so you may want to check and see if a newer version is available. Radio Mobile is based upon the Longley-Rice propagation model, and adds several morphological correction factors which can be used to change the model to more closely predict the actual propagation in an area based upon land use and terrain. As shown in Figure 4-4, accurately setting these morphological parameters is critical. The best way to determine the right parameters is by doing site surveys in areas similar to those you wish to cover. Once this real-world data is available, it can be compared to the modeled results, and the morphological parameters associated with the different areas can be "tweaked" until the predicted behavior closely approximates the measured data. To do this, look at area averages and try to get them to match as closely as possible. Remember the model does not have actual morphology, so it does not know where individual roads, trees, and buildings are located. Because of this the model cannot accurately predict local shadowing caused by these objects. The best it can do is predict the average propagation behavior in an area based upon an average morphological density. Therefore you should not expect absolute accuracy of predictions against measurements. The accuracy of the model increases the further you get out of local clutter. If base station and far end equipment is located above the level of local buildings and foliage, the model will predict more accurately because the clutter is now nothing more than a source of multipath. The closer to the ground you place one of the stations, the more local morphological features will begin to impact the accuracy because of the local shadowing they generate.

Nonetheless, modeling is a valuable tool. Though it may not be able to tell all, it can often tell enough about the propagation in an area to give a level of comfort about what, on average, to expect as the coverage provided by a base station. This can be useful for ranking locations, or simply checking to see if the coverage of a site appears sufficient to justify its costs.

The first thing to do with this or any propagation prediction software is to gather the appropriate terrain and morphological databases that the model needs for prediction. Also gather any digital street maps that you may want to use as base maps for plotting your coverage on. These tasks are straightforward when using Radio Mobile, because the program is designed to access publicly available Internet databases containing terrain data as well as street maps.

The next requirement is to know the RF performance characteristics of your equipment. These include frequency of operation, RF power output, receive sensitivity, and antenna characteristics for both the base station and the client end equipment. For this example I'll use the following RF characteristics, which are similar to those used for unlicensed 802.11b equipment operated in a WISP system:

- Frequency = 2450 MHz
- Base Station Tx power = 0.1W
- Base Station Rx sensitivity = −94 dBm
- Base Station Antenna Pattern = 360 degrees
- Base Station Antenna Gain = 14 dBi
- Base Station antenna azimuth = 0 degrees
- Base Station Antenna Height = 60 feet
- Cable loss = 4 dB
- Far end Tx Power = 0.1W
- Far end Rx Sensitivity = −94 dBm
- Far end Antenna pattern = 30 degrees
- Far end Antenna gain = 5 dBi
- Far end Antenna Height = 15 feet
- Cable loss = 0 dB

The final requirement is an estimation of the morphological correction factors to be added to the model. These can be determined by site survey, or by estimation from known propagation behavior in similar areas.

With the above information loaded into the model and applied to terrain data only it generates the plot in Figure 4-4a. This is probably an accurate representation of coverage in an open desert or coastal plain, but since populated areas do not have this open, terrain only characteristic, additional losses due to the morphology of the area must be considered. Adding loss associated with forested morphology to the terrain yields the plot shown in Figure 4-4b. As you see this significantly reduces the predicted coverage, as would be expected when trying to cover an area with dense foliage at these frequencies. Figure 4-4c shows the predicted behavior when the model is tuned for the morphology associated with a typical newly developed suburban subdivision, which is the actual environment being covered by this site. The correction factors selected were based upon data collected in this and similar environments, and provide a reasonably accurate estimation of the sites real-world behavior. This illustration shows the importance of accurate terrain and morphological data. It also shows why you should do some field-testing to validate predictions before blindly believing the output.

Gathering accurate field measurements is a task that can be accomplished in association with collecting other information about the area where the system will be deployed. It can be done as part of a comprehensive site survey.

The Comprehensive Site Survey Process

Since we are dealing with a radio-based wireless technology, it exhibits the irregular propagation characteristics of an RF-based service. As discussed in previous chapters, the RF signal is subject to fading, multipath, and many attenuation variables along its propagation path. These variables cause both the covered area and the spot coverage within the covered area to be difficult to predict accurately. One way to determine coverage accurately is by performing a site survey. The site survey involves the temporary installation of equipment and the use of measurement tools to actually measure the signal in the desired coverage area.

Since the RF site survey requires such a significant effort, it should be conducted as part of a comprehensive site survey. A comprehensive survey will consider many factors that need to be known in order to deploy a system that meets the coverage, capacity and cost requirements set out for the network. In addition, the site survey can be an invaluable tool in determining what needs and limitations (like availability of power and network connectivity) exist vis-à-vis the system

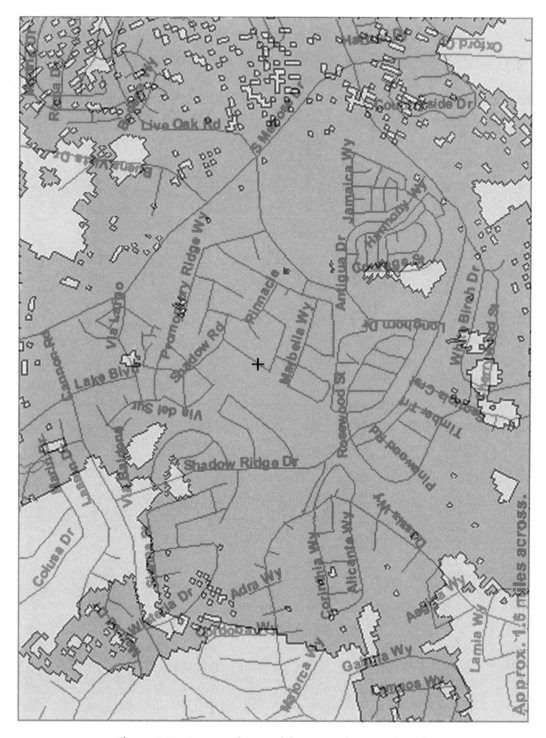

Figure 4-4a: Propagation model output plots terrain only

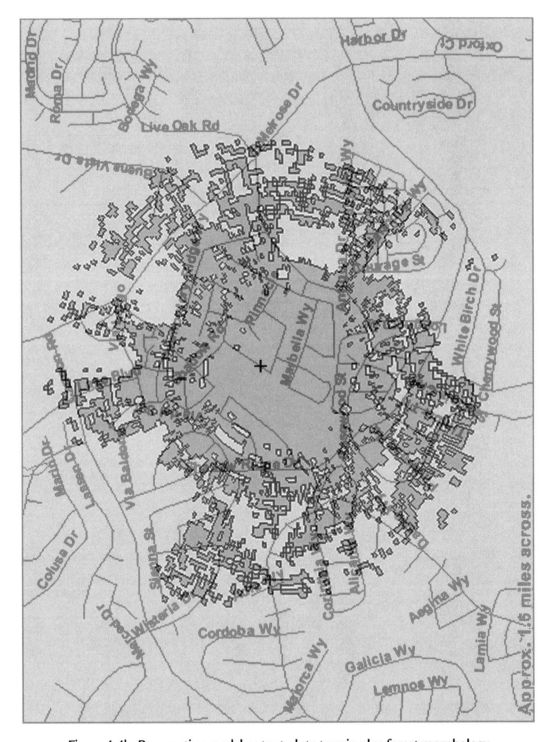

Figure 4-4b: Propagation model output plots terrain plus forest morphology

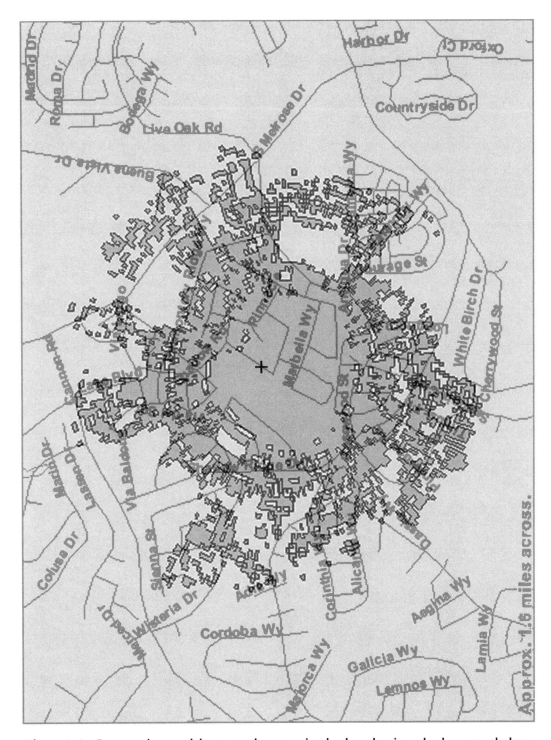

Figure 4-4c: Propagation model output plots terrain plus low density suburban morphology

installation. The following site survey outline identifies the issues that need to be addressed during the survey.

Survey Activity Outline

- Location Identification
  - Latitude/Longitude
  - Location Address
  - Owner/Manager Contact Information
  - Structure Type
    - Structural Material
  - Area of desired coverage
- Identification of customer requirements
  - Coverage
  - Capacity
  - Security
- Identification of RF Zones
  - Based on Desired Coverage
  - Based on desired capacity
  - Based on interference management
  - Based on power and interconnect availability
- Equipment and Technology Selection
- Spectrum Analysis
  - Identify existing interferers
- RF Survey
  - Measure coverage from test locations
  - Plot actual vs. desired coverage in area
- Network design
  - Interconnect to radios
  - Interconnect to rest of network
  - Capacity
  - Security
  - Access control
  - IP addressing

- Equipment Selection
  - Transceivers
    - Make
    - Model
    - Frequency
    - Vendor
  - Antenna
    - Make
    - Model
    - Pattern
    - Vendor
  - Mounting Equipment
    - Type / Description
    - Vendor
  - Network
    - Interconnect
- Cable
- Radio
    - Routers
    - Switches
    - Hubs
    - Access control
- Conceptual Design
  - RF
  - Network
  - Interconnect
    - RF site interconnect
    - Network interconnect
      - ◆ To internal data network
      - ◆ To Internet
  - Costs
    - Equipment

    — Labor

    — Recurring

■ Facilities lease

   • Availability and cost

■ Site lease

   • Availability and cost

Identification of Requirements

Before equipment can be placed and measurements taken, it is important to determine what is expected of the network and how it will be used. The area to be covered, the number of users, and the services used all have an intertwined relationship. Understanding them will be of great help in determining the best equipment solution and locations to effectively serve the users needs.

For example, the design requirements of a system to provide Internet access to a 20' by 20' coffee shop and provide service to five simultaneous users is significantly different from a system designed to cover an entire 20,000 square foot office area and all the computer users in it. Differing even more are the requirements of a Wireless ISP (WISP) that wants to provide ISP services to a town or community.

In the first case above the room is small and the user community is also small. It is realistic to expect that a single-access point located within the room will provide sufficient coverage and capacity. The outcome of the site survey in this case is determination of the best location for the access point based upon RF coverage and ease of getting power and Ethernet cabling to that location. The survey should also identify what other hardware, like a router or gateway, is required to implement the service.

The second case is much more complex. How is the office space laid out? What construction materials were used in interior walls, ceilings and floors? How are the users distributed? What concerns does the customer have regarding signal leakage out of the building? What cost does the customer have in mind to provide this solution?

All of these issues will affect not only the RF issues surrounding the location, power level, antenna configuration, and channel reuse of the system, but also the network requirements surrounding traffic segmentation, security, routing, and switching.

The WISP case has different complexities. What are CAPEX, OPEX, and revenue expectations for the system? What are the reliability expectations? How big an area is to be covered? What are the locations of available sites in which to install the access point or similar equipment? Does sufficient Internet access bandwidth exist at these sites? If more than one site is needed for coverage, what are the available backhaul methods? What is the terrain of the area? What is the morphology of the area? How many users will be in the covered area? What will their usage patterns be like? How will the users be distributed across the coverage area? How will the system provide access control? What security concerns does the WISP have? Will the user have Customer Premise Equipment (CPE) located outside their home, or is the service expected to provide RF coverage to CPE located within the residence?

As you can see, the larger the system deployment, the more complex it becomes. Knowing what is expected of the system is the first step in resolving the best method of either meeting those expectations or setting new more realistic expectations.

Identification of Equipment Requirements

After gaining an understanding of the customer's needs and expectations, the implementer should be able to select equipment that best suits the needs of the environment. For example, a cheap consumer quality 802.11b AP may be the perfect solution for the coffee shop, because it is inexpensive and includes a low-end router, DHCP server, and provides NAT functionality. Thus it is a one-box solution for this particular environment.

This consumer grade AP solution would be a poor choice for the office or WISP examples. The office solution has need of an AP with features like remote management capability, power control, the ability to use external antennas to customize the area covered by each AP, the best encryption available, and the ability to be upgraded with new firmware so it can offer state of the art capabilities for the longest period of time.

The WISP, on the other hand is probably not best served by a traditional 802.11b AP. The requirement of large area coverage from minimum locations means that the equipment will need to be tailored to high EIRP devices with high gain antennas. Moreover, the equipment will be mounted outdoors, thus requiring weatherproofing to be a design consideration. There are numerous companies who offer such products (Motorola, Alvarion, Proxim, Navini, and Vivato to name a few) and new ones seem to enter the market every month. Some of these solutions are 802.11 compatible, others use proprietary air interface solutions.

Each solution has its place in the market. Understanding the requirements of the system will assist in selecting the manufacturer and solution that is best suited to serving those needs.

For the remainder of this chapter, it is assumed that a working knowledge of the capabilities of the equipment being contemplated already exists. If it does not, then one should become familiar with the capabilities and expected coverage of the equipment before embarking on the site survey. Many of the techniques outlined below can be utilized to determine the coverage and capabilities of equipment, and can be used to evaluate the equipment in a known environment.

The Physical Site Survey

Once system requirements are understood, a physical site survey can commence. Obtain as much existing information as possible. Items like topographic maps, satellite images, building blueprints, and so forth will be invaluable in planning the survey.

With these documents in hand, you can begin to physically survey the property. Walk or drive around the area to be covered to get a visual understanding of the area to be covered, noting any major obstacles to coverage.

If outdoor coverage is planned, one should look for and note dense trees, buildings, and hills between the radio site and the desired service area. Note how far away you can physically see the radio site from as many locations within the desired service area as possible.

For indoor systems note the location of metal or cement walls and floors, as well as the location of large metal objects like refrigerators. Also note the location of "utility walls," i.e., those walls that contain dense runs of piping and or electrical cables.

Determine where the equipment needs to obtain its data and power connections. If the survey is of an office building, and the equipment needs to be connected to an existing computer network, note where this network equipment is located, and how new cables will need to be routed to get there.

If remote connections are needed, in other words the connection to a data source does not reside on the same site being surveyed, note where the telco facility room is located on the property and where the other end of the connectivity must go. Also note how cables or wireless facilities can be routed from a central point of interconnect to the radio site locations, and whether there is a secure space where the network and interconnect hardware can be located.

Determination of Antenna Locations

Determining optimal antenna locations is the key to a successful deployment. An optimal location serves a multitude of needs: it provides optimal RF coverage; meaning it can be optimized to provide sufficient coverage of the area without leading to significant interference elsewhere in the system, it has easy access to power, it has easy access to network interconnect facilities, it can be easily installed and secured, and it has reasonable access for future service needs.

Since 802.11 hardware is easily available and has a large base of testing tools, I'll use 802.11 as the basic technology to discuss the decisions and tools required for system design. Even if the system you are designing is not 802.11-based, you can use the same procedures and criteria in designing a network based on 802.16, 802.20, or any other standard or proprietary solution.

The first step is to select an equipment solution based upon the needs of the customer and the environment to be covered. Select solutions that will most easily or most cost effectively meet the coverage and capacity requirements of the area.

Once the equipment is selected you have a baseline for the RF transmit power, receive sensitivity, and antenna options. As previously discussed, these numbers are used to determine the available path loss using the following equation:

$$L \, [dB] = P_{tx} \, [dBm] + G_{tx} \, [dBi] - P_{rx} \, [dBm] + G_{rx} \, [dBi] - M \, [dB] - C_a \, [dB]$$

Where L is the link budget in dB, P_{tx} is transmit power, P_{rx} is receiver sensitivity, M is fading margin, C_a is the attenuation of area construction material, and G_{tx} and G_{rx} are antenna gains on the transmit side and receive side respectively.

Using the conservative power levels and antenna gains associated with common AP equipment used for indoor office LAN type deployments yields the following:

$$L = 15\ dBm + 0\ dBi - (-8\ 2\ dBm) + 0\ dBi - 10\ dB - 8\ db = 79\ DB$$

Using the graph from Figure 3-1 (Chapter 3), you can see that in an office type environment where the propagation will consist of some line of sight and some non line of sight paths, the expected coverage of a single AP location could range from 60 to 150 meters depending on the actual conditions of the path. If no interior walls block the path, the signal will propagate further. High-density walls will attenuate the signal more severely.

Remember, this is a simple graph, and does not take into account all the propagation variables that will be found in the field. The distances derived from the graph are average numbers. There will be areas of a building (like central cores and elevator shafts in a high rise building) that exhibit far greater attenuation than the average in the office environment. This is why a site survey is helpful: it allows you to measure the actual propagation environment so you can decide precisely where radio sites should be located in order to provide best coverage, capacity and interference management.

Depending on the physical layout of the space to be covered and the availability of power and interconnect, a number of location options are viable: you could use an omni antenna located on the ceiling in a central location, or you could use a directional antenna located high up in a corner or along an outside wall and pointing toward the space to be covered.

Possible solutions are shown in Figure 4-5:

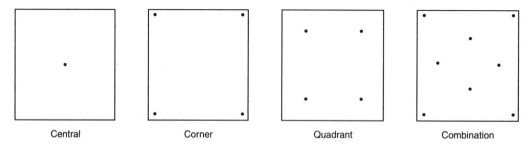

| Central | Corner | Quadrant | Combination |

Figure 4-5: Base station location option diagrams

Real buildings may not be square or rectangular, however the same location opportunities exist regardless of the building's shape. As shown in Figure 4-6, one method of breaking up a simple or complex floor plan is to grid it into squares that approximate the coverage you expect from each base station. By gridding the floor, you get a sense of how the space is organized, and you can analyze the user density within the grid. If you find too many users in a square, break it down further so you can see the actual area that needs to be served by each base station in order to accommodate coverage and capacity needs. With the area so divided, the selection of technology and the ideal placement of base station equipment can become much clearer.

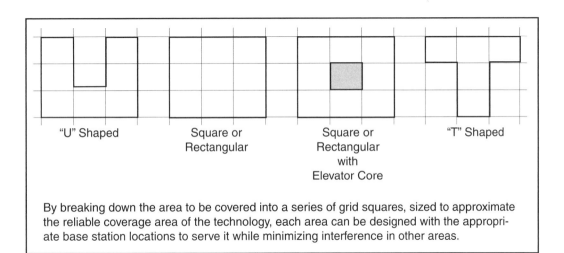

| "U" Shaped | Square or Rectangular | Square or Rectangular with Elevator Core | "T" Shaped |

By breaking down the area to be covered into a series of grid squares, sized to approximate the reliable coverage area of the technology, each area can be designed with the appropriate base station locations to serve it while minimizing interference in other areas.

Figure 4-6: Building layouts and gridding

RF Site Survey Tools

Now that possible AP or transmitter site locations have been identified, it's almost time to do some testing and measurement of the signal strength, noise, and interference in the environment. Before you can begin measuring these values, you need to acquire test equipment to do the measurement and data collection. Luckily such hardware and software is readily available and in some cases free.

If you are using proprietary hardware it will be up to the equipment provider to supply the measurement software and procedures. If you are using a standard 802.11 solution there are numerous software choices.

There are three classes of measurement and test software available for 802.11 RF testing and monitoring, each having its own benefits and limitations. The first class is the client manager that is included with most client cards. It is the simplest tool and has few or no features. Some can display the signal and SNR of the access point to which you are connected, others only show a bar graph of signal strength and the speed at which you are connected to the AP. Still others, like the Lucent Orinoco client manager shown below, can show all active channels in the vicinity so long as they are associated with the same network and have the same SSID In addition they show the MAC address of the AP, the signal strength, noise level, and SNR. The Orinoco client manager also has rudimentary data logging capability. It can save the measurement results on 1 second or greater intervals automatically, or it can be set up for manual logging where the user must tell it to log a measurement and provide a text explanation to go with the measurement. This manual mode can be useful for taking "waypoint" measurements, in other words measurements that are correlated to a known point in space.

This class of software can be useful in small-scale site surveys, like the coffee shop example mentioned earlier, or to use in spot-checking coverage in a larger deployment. It can also be a useful troubleshooting tool because each user of the network will have this software installed in their computer when the wireless card is installed. A first echelon of troubleshooting user problems would be to have the user open the client manager program and look at the information displayed.

The next class of software contains the free solutions like Kismet if you use Linux, or Netstumbler if you use Windows. Both have some limitation vis-à-vis the client cards and GPS formats they support, so care must be taken to assure compatibility with the rest of the test setup. The big benefit of these software packages is their improved feature set. They can be used to monitor all AP activity on all channels simultaneously. In addition they have a GPS interface, which makes them much more useful if outside measurements are contemplated. They log data to a file, and have the ability to export these files in a number of formats for post-processing and analysis.

The final class of software is the commercial package like Airopeek, AirMagnet, and Ekahau Site Survey. These packages are significantly more functional than the freeware packages. They also have compatibility issues with the Client Cards and

GPS formats they work with. They are also expensive: $1000 to $2500 per copy of the software. It is also worthy of note that they have all been designed with certain purposes in mind, and they do not have 100% overlap of capabilities.

For example all of the above solutions are capable of collecting data, but only the Ekahau Site Survey software has the built in ability to create coverage maps directly from the software to a map image. To map the output of the other packages requires exporting the information and manually creating a coverage map with another software package.

There will continue to be new developments in the field of software and hardware for site surveying, monitoring and evaluation. It is well worth your time to search out currently available options. Evaluate several choices, and select the one that seems to best fit your particular requirements.

The Site Survey Checklist

Before you head out on your survey, take the time to assure that you have all the common items you may need on site.

The obvious items are such things as:

- The selected radio hardware solution
- Your portable computer configured for measurements
  - Computer
  - Client card matching the chosen hardware solution
  - Measurement software
  - GPS
  - A cart or sling to carry the computer
  - Extra batteries and a battery charger/AC adapter
  - Any cables needed to connect external devices
- Spectrum analyzer

Less obvious items include:

■ Mounting hardware for temporarily installing the radio equipment

■ Tools to accomplish the temporary installation

■ Extension cords to reach power

■ Network cables to reach existing network

■ Duct tape to tape down these cables

■ Wire, tie-wraps

■ Antennas appropriate to the initial design analysis

■ High-quality coax jumpers to connect the antenna to the hardware

■ Stepladder

The RF Survey

The survey is accomplished by temporarily installing the selected hardware solution at one or more of the predetermined locations, powering it up, getting it configured and operational, then using a client device and special software to collect information on signal strength, noise and SNR ratio.

Determine the best way to mount the hardware temporarily in the locations you've predetermined from studying the area to be covered. You want it secure, but do not want to permanently damage the area where you are mounting the hardware.

Once mounted, power it up and perform any configuration necessary to get it operating. Turn on your survey device and look for the signal from the hardware you just installed. If you are very close (within 10 feet) to the hardware you should see signal strengths ranging from –40 to –60 dBm. If you see appropriate signal strength from the desired equipment, you are ready to begin surveying the coverage area.

It is important to remember that 802.11 as well as any number of other technologies operate in unlicensed spectrum allocations. If the technology you are deploying is operating under FCC part 15 rules, a few initial tests are in order. First, use a spectrum analyzer to look for existing carriers in the band. Because Part 15 devices use different modulations, the only way to see and characterize the use of the band is to look at the spectrum analyzer plot and identify all carriers occupying the band. Next, check your survey software to see if it has identified any other equipment using the same standard as your equipment working on or near the channel selected for your equipment. If you see other operating hardware, make sure you set your equipment to operate on a nonconflicting channel. Also check the noise floor on the chosen channel to assure it's below −90 dBm. If the noise floor is over −90 dBm, there is a good possibility that another device using noncompatible modulation is operating on the channel. Because this noncompatible system will be seen as noise or interference by the new network, it is best that this channel also be avoided at this location in order to assure the best coverage and capacity from your device.

Now that you have the equipment functioning on a clear channel the RF survey can commence. Begin by moving around the desired coverage area and noting the signal strength and SNR at as many locations as possible. This is where your environment and selection of measurement software becomes critical. If you are measuring outside, GPS can be used for positioning, and with compatible measurement software GPS can be used to log location and signal strength and SNR at that location. Without GPS you will have to manually log as many points as feasible, as well as keep accurate track of your path. With the manually logged points and knowledge of how you got from point-to-point, you can manually create a coverage map from the information collected. Measuring indoors presents a situation similar to having no GPS. Since GPS does not generally work indoors, it cannot be relied upon for positioning in the indoor environment.

Continue to move away from the test node until the signal falls below noise level. Move back into the coverage area until you again acquire signal. If you have measurement software that is capable of collecting location data, take advantage of this capability and move randomly around the periphery in and out of coverage. The software should collect sufficient data to define the site boundary. Now move randomly inside this boundary, collecting as much data as feasible in the area covered by the test location. Use your mapping/plotting software to generate a coverage map of the area for further review and analysis.

If you do not have the ability to collect accurate positional data with your software, try the following procedure. Keeping the signal 1 to 3 dB above the noise floor (SNR = 1 to 3 dB) move around the entire periphery of the covered area. As you will begin to notice, there may be significant changes in the location of this outer periphery. You may in some cases notice that a movement of 5 feet at the periphery requires you to move 20 feet closer to the test site in order to maintain the signal, in other areas the opposite will be true. Continue to move about the periphery and accurately draw this contour on a map, picture, or blueprint of the site.

You will now have a map defining the limits of coverage of the test location. This is not the same as the useable limit of the site, defined as an area with sufficient SNR and fade margin to provide solid connectivity to the user, but does define the interference limit of the location. This will become important in considering the positioning of other RF locations and the channel selection for these locations.

Now repeat the measurement process with new SNR levels. 5 dB, 10 dB, and 15 dB levels are reasonable starting points. These contours should be plotted on the same map as the first measurement.

Perform a quick evaluation of the data you've collected. An example evaluation is provided by Figure 4-7. Does the test site cover the desired area? Are there any coverage holes in critical areas? If the coverage is not as expected, or there are critical coverage gaps, try to identify why the test site is not behaving as anticipated. Look carefully at the coverage contours; is there a clearly identifiable shadow in the coverage? If so, there is most likely a construction anomaly or other obstruction in the path. Having identified the location of this blockage, determine if it can be avoided by moving the test site to a new location that avoids the obstacle and perform the survey again. If, for some reason the test site cannot be moved, then the coverage holes will have to be accepted as areas of poor coverage by the system, or they may be correctable with a signal repeater located in the weak area.

Once you are satisfied with the coverage, repeat the above procedure on all remaining test locations.

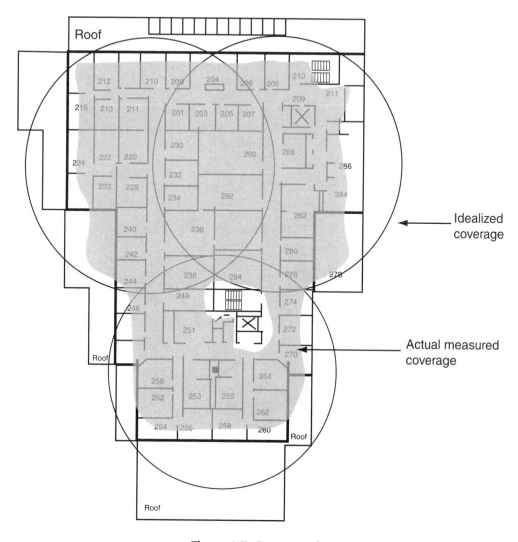

Figure 4-7: Coverage plot

Data Analysis

With the data collected and visually plotted on a map or other image representative of the area to be covered, numerous details will become evident. These details will be useful in finalizing a system design that best meets the needs of the customer.

Because the analysis, and the changes necessary to conform the network to real-world needs is an iterative process, the review is best conducted utilizing a flowchart methodology. The flowcharts in Figure 4-8 are representative of the

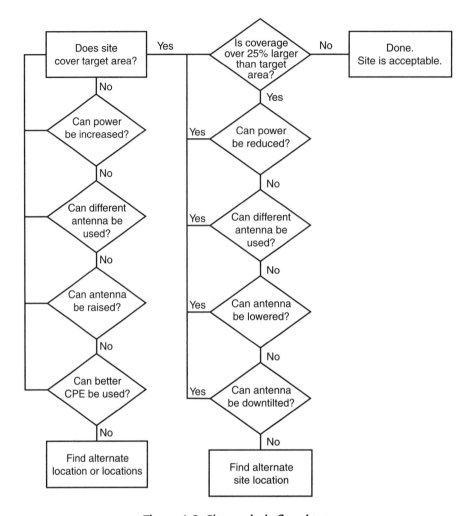

Figure 4-8: Site analysis flowchart

approach used to analyzing the data and make changes as needed to conform the solution to the real world as characterized by the survey. The first review should be conducted while the survey equipment is still mounted and operational. Upon seeing the initial survey results, you may decide that some optimization needs to take place. It's easier to do the additional tests now, rather than try to recreate the survey installation a second time.

The first analysis should be to review the coverage of the system. The first flow-chart is used to analyze whether the site(s) provide coverage to the intended area, and if not, offer a number of alternatives to correct the coverage.

Once coverage is deemed acceptable, the second flowchart is used for managing excess coverage. Excess coverage is a problem on several fronts: Depending on how big the coverage extension is, it may be a security issue. For example if the wireless application is to provide connectivity for an office LAN, and the coverage of the WLAN extends outside the building into the parking lot or into the street, there is an opportunity for unauthorized access or monitoring of the WLAN and all the traffic it contains. Such an unintended coverage extension should be addressed first by minimizing the unintended excursion. If it cannot be completely eliminated, then additional security measures may be desirable on the WLAN. The second problem generated by excess coverage is overlap into the primary coverage area of another radio site. Because the client card normally identifies its primary radio site based on signal strength, there will be areas in the overlap where there is contention for the strongest signal. Fading and multipath exacerbate this problem, and lead to moment-by-moment changes in signal strength between the contending signals. The result of this can be the inability of the client to maintain communication with a single radio site. The client may "bounce" between competing signals at random, leading to throughput issues, or in some cases complete loss of data or connectivity.

More importantly from a system design standpoint, this excess coverage leads to interference with other radio sites using the same channel. This issue is critical in deployments using multiple radio sites, since there are a limited number of channels. This limited channel set will need to be reused by the radio sites over and over again within the coverage area. Interference is avoided by physical separation of co-channel reuse sites (see Chapter 3).

Power reduction, antenna selection, antenna downtilt, and site placement can all have a profound effect on controlling coverage of a site. The first corrective action to reduce the coverage should be power adjustment. In order to select an appropriate power level you must refer to the link margin used when starting the survey exercise. The published receiver sensitivity plus a fade margin was used in the link budget as the base signal strength necessary to maintain a communication link at the desired throughput. Using this value (receiver sensitivity plus fade margin), reduce the power of the radio site until the edge of the desired coverage area is provided signal at this level. This is a straightforward process utilizing the data you've already collected. Your coverage map already shows signal strength across the coverage area. Look at the measured signal strength at the edge of the defined coverage area, and subtract the measured signal from the required signal. The result will be the number of dB the transmitter can be power reduced.

If the coverage area can be conformed with a power reduction, next make sure that the power reduction has not created any coverage holes inside the desired coverage area. Do this by identifying any areas on the survey map that show shadows or weak coverage. Subtract the number of dB you've reduced the power, and ensure that these weak areas still have sufficient signal to meet expectations. If they do not, then you could either increase the power until they do, consider adding an external antenna to the clients in the weak area, or add a signal repeater to the weak area.

You should keep in mind that no RF-based system is perfect. Even a well-designed system has coverage gaps. The best that can be expected is for the coverage to be useable over most (85 to 90%) of the covered area.

If power alone is insufficient to reduce the covered area, using a lower gain antenna, or a directional antenna downtilted toward the center of the desired coverage area may solve the problem. Lowering the antenna placement may also help. Unfortunately, all of these solutions will probably require additional survey time, since the previously collected data cannot be easily utilized to analyze changes of this magnitude.

Now that your design has been through enough iteration to assure maximized coverage inside the desired coverage area, and minimized coverage outside that area, it's time to select channels.

Regardless of the technology selected, there will be a limited number of channels available for use. The first limitation on available channels will be those assigned by the Government regulator in charge of spectrum allocation; other users of the spectrum in the area will cause the second limitation. In the case of technology using unlicensed spectrum (such as 802.11 products), the available channels might be used by devices as different as cordless phones and video transmitters.

If the system you are constructing has fewer transmitter locations than available channels, the deployment is simple: just assign unique available channels to each transmitter. If the number of locations exceeds the number of channels, then a frequency reuse plan (as discussed in Chapter 3) will need to be designed and implemented. The coverage maps generated during the survey and corrected for power level are of great value in accomplishing this task.

The background provided in this chapter becomes the basis for deploying effective networks. Chapter 5 will begin to utilize the tools and techniques we've discussed and show how they can be used not only to determine coverage, but to identify alternative system designs and compare and contrast them. Through this process of evaluating alternatives, the system with best compromise between coverage, capacity, utility and cost can be identified.

CHAPTER 5

System Planning

- System Design Overview
- Location and Real Estate Considerations
- System Selection Based Upon User Needs
- Identification of Equipment Requirements
- Identification of Equipment Locations
- Channel Allocation, Signal-to-Interference, and Reuse Planning
- Network Interconnect and Point-to-Point Radio Solutions
- Costs
- The Five C's of System Planning

System Planning

Now that we have reviewed the basics of radio operation, propagation, and predictive and actual performance measurements, it's time to see how this information is used as part of the design criteria in a system to actually provide services to a customer base.

System design must consider far more than just the RF aspects of the system. If the system is to function optimally and be cost effective, such diverse topics as equipment selection, real estate, construction, interconnect, power, and maintenance must be considered. Each of these topics has an initial capital cost and, with the exception of construction, an ongoing cost.

System Design Overview

Because of the myriad interactions you will encounter in designing a system, a flowchart is helpful for identifying the selection criteria for each of the key aspects of system design. Because there are so many different unique business opportunities that can be served with wireless data systems, it's impossible to review them all in this book. Instead I'll look at three distinctly different models, and walk through a design exercise for each. The first example system will be a single AP "hotspot" or small office LAN. The second example will be a far more complex MultiAP office LAN or "hotzone" requiring frequency reuse in its implementation. The third system will be a Wireless ISP (WISP) type system that is expected to cover a large outdoor area and provide Internet connectivity to a large, geographically dispersed user base. The WISP system could be composed of a single site covering a small town, or potentially hundreds or thousands of sites covering multiple counties or MSAs. Fundamentally they are all the same, though the complexities and need for managed spectrum grow with the size of the deployment.

Regardless of the scale of the system being deployed, there are a number of individual activities that have interaction with each other. For example, selecting locations for installing the radio hardware will be influenced by cost, coverage, and capacity needs of the system. Cost, coverage, and capacity are influenced by the selection of radio hardware and the frequency of operation. So, you can begin to imagine the complexity involved with the design of a large system. Each individual topic surrounding system design has its own associated flowchart which identifies activities and go/no go decision points. As well, each flow chart must consider other parallel activities occurring under another separate topic, so that you assure that decisions and compromises made in the pursuit of one area of design do not negatively impact system viability because they ignored key factors of a separate decision matrix that they affected.

To simplify the overall decision matrix, I'll present individual flowcharts for each key activity. These flowcharts will show precursor or parallel activities that will need to be considered or reviewed when making final decisions surrounding individual key activities. After discussion of all the planning criteria, I'll show how these factors apply to real-world systems by using the flowcharts and decision matrices to plan actual systems, and show some of the trade-offs.

Location and Real Estate Considerations

Of course, the first thing you need to know is where the system will be deployed, what it needs to cover, and how much capacity is needed. In a hotspot or office LAN system, a physical address and suite number are necessary. In addition, a floor plan or other identifying criteria showing the area(s) to be covered should be acquired. Also discover as much about the property as possible. Blueprints, building drawings or other documentation concerning of the type of construction present in the building will be useful in the exercise of estimating coverage. Another key bit of information will be the name and contact information of the building owner, property manager, or other entity that may require coordination or approval of work on the premise.

If the system is of the WISP variety, there are additional needs. Since a building no longer defines the coverage, the physical area to be covered should be identified on a map or area image. This area should be inspected for available towers or multistory buildings that could be used to locate equipment and antennas. Latitude/longitude and height of antenna mounting locations of these buildings or towers should be identified. In addition, it is important to ascertain who is the owner or property manager of target buildings or towers.

You should also remember that many jurisdictions use zoning and permit processes for any communication facility, regardless of whether it uses licensed or unlicensed spectrum. It is critical to discover and comply with any local zoning or building and safety requirements early in your planning process. Failure to do so may lead to significant delays in deployment or, worst case, the local jurisdiction fining you and forcing you to cease operations and remove the equipment.

Because of the area to be covered, numerous possible equipment locations will exist. Determining where to concentrate your efforts requires a rapid assessment of which properties are best from an RF design standpoint. Assuming you have the specific operating characteristics of your equipment, the use of a propagation-modeling tool can prove valuable for assessing coverage from each of the location options.

In order to use such a tool, the parameters of the system need to be known, and assumptions need to be made about the conditions present at the CPE location. For example, if the CPE is to be located indoors near the user's computer, additional path loss due to the construction of the building in which the CPE resides must be considered. If the CPE can be mounted outside, clear of local obstacles, then the propagation losses will be significantly lower, and the site's coverage greater.

As examples of this, the propagation plots in Figure 5-1 were computer generated using the Longley-Rice propagation Model. The plots are based upon the same transmit power output and receive sensitivity. Only antenna gain and placement at the CPE has been changed. Figure 5-1a shows the coverage achieved from the base station or access point (the terms base station and access point can be and are used interchangeably. While the term access point was once unique to 802.11 hardware, it can now be seen referring to any number of base station products supporting wireless data) to a CPE unit using a 0 dBi gain omni antenna at street level, while Figure 5-1b shows the coverage achieved from the same base station to CPE using a 15 dBi antenna mounted at 15 feet elevation (sufficient to clear the roof of a one story home).

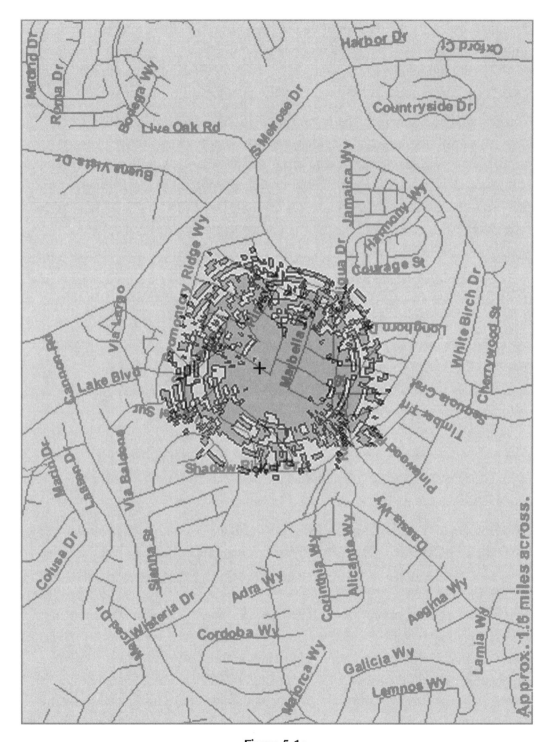

Figure 5-1a

Figure 5-1b

The Longley-Rice model predicts long-term median transmission loss over irregular terrain relative to free-space transmission loss. The model was designed for frequencies between 20 MHz and 40 GHz and for path lengths between 0.1 km and 2000 km.

The Longley-Rice model is used in a number of commercial propagation models that are utilized for analyzing signal propagation in such commercial applications as cellular and PCS communication systems. The model is well known, and accepted by the FCC as a method of predicting coverage of broadcast facilities.

The model uses actual terrain data and predicts the median signal strength across this terrain based upon a combination of distance and terrain obstacles. The model also has a generalized morphology (land use) attenuation factor that can be attached to each site. This allows additional attenuation to be added to accommodate such things as building density and foliage.

While the coverage plots in this book show coverage by the use of a grey tone to indicate coverage area, the actual plots generated by the software are in full color and consist of a number of colors, ranging from green to red. Each shade is associated with a 3 dB range of signal strength, with greens being high signal strength and red being low signal strength.

To use the prediction plot to design a system, you must determine three factors:

- How much fade margin does the system need?
- How much building attenuation must be overcome?
- Will the client device be externally mounted with a high gain antenna?

Since radio propagation is continually effected by multipath, the signal is always in a state of flux. This flux is known as fading. Even a stationary transmitter and receiver will see path fade between them based upon objects like trucks, cars and people moving in the environment. It is good engineering practice to use 8 to 10 dB of fade margin in a system design.

If the signal is to be received indoors, the building itself becomes an additional source of attenuation. This can range from 5 to 7 dB for wood frame construction to over 25 dB for office buildings with metallized glass facades.

Finally, the above factors need to be subtracted from the baseline receive sensitivity of the client device. In an 802.11b system, receive sensitivity on a high quality card ranges from –92 dBm for 1 Mbps throughput to –83 dBm for 11 Mbps throughput.

If the client device can make use of a high gain directional antenna, then this gain can be added to the path loss, or to make it simple, just add it to the receive sensitivity. So with a 15 dB gain Yagi antenna the –92 dBm sensitivity increases to (–92 dBm –15 dB) or –107 dBm. You have not really increased the receive sensitivity, but you have added gain to the receive chain, which for all intents accomplishes the same thing.

So, a reliable communication link to a base client card at 1 Mbps requires a consistent –92 dB signal. Because of the fading nature of radio waves, it is necessary to add the fade margin to the –92 dBm minimum signal. This means that the signal required for a reliable link will need to be (–92+10) or –82 dBm. This signal level is sufficient to offer outdoor communication, but offers no margin to penetrate buildings. Additional signal is required for this. Again, 5 to 10 dB of additional signal strength will be necessary to penetrate light construction, so the required signal strength rises another 10 dB to –72 dBm.

If the client device makes use of a 15 dB gain antenna located outside above roofline, only the fade margin needs to be considered. Thus, a signal of (–92 dBm – 15 dB + 10 dB) or –97 dBm is required for a reliable communication path.

Looking at the contours and the legend, you can identify the areas that will be served with signals of the strength discussed above. Those points that show sufficient signal strength to meet the minimum requirements will generally be capable of supporting a reliable communications link.

Using the modeling tool can allow rapid review of a number of candidate radio site locations, and allow you to select the optimal locations for providing coverage to the target area.

As a location is identified, a propagation model should be run with the specifics of that location such as longitude, latitude, and height. This will give you a way to evaluate the anticipated coverage of each location and quickly identify which properties are ideal candidates and which are poor choices. By using the model for first pass identification of the best sites, you can focus your efforts on those sites first.

Another key consideration in the real estate realm is the availability of space, power, and interconnect for your system. Often the ideal RF location may be lacking one or more or these elements. Determining what additional costs and complexities are associated with getting the required space, power, and interconnect to the desired location is a critical part of the selection process. The flowchart in Figure 5-2 identifies the critical aspects that should be considered when selecting a site. As with most everything else in a radio-based system, there will rarely be a perfect solution, so compromises on certain factors related to sites will need to be made. For example, the site may not provide optimal coverage, but may be the only site in the area for which a zoning variance can be obtained. As shown in the priority of the flow-chart, coverage should be the prime consideration, although often other considerations will dictate the use of a site that is an imperfect solution from a coverage standpoint. The flowchart is best utilized for analyzing a number of different site options. The trade-offs associated with each can then be used to make the appropriate business decisions about the best location that has been analyzed.

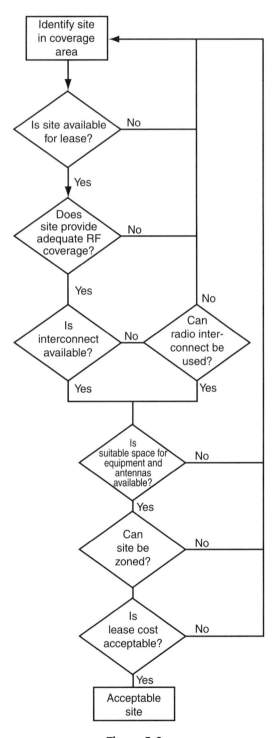

Figure 5-2

System Selection Based Upon User Needs

The next considerations have to do with subscriber behavior. What form factor is acceptable for CPE? What average capacity requirements and usage characteristics will each user have? What security level will the user expect? And, finally, what is the acceptable initial and ongoing cost of the service?

These considerations are primarily driven by the needs of the customer. Meeting them will affect equipment and real estate decisions. For example, in a hotspot the equipment decision might be driven by the following logic: a hotspot serves customers that are using laptop computers or PDAs equipped with wireless access. The hotspot provides these users with short term Internet access. The system is primarily used for web browsing and email type activities. In this case the average use per customer is low, and the connection to the Internet is probably far slower than the speed of the radio interface, so the radio system is not the capacity bottleneck. In addition, the system must rely on the fact that the user is already equipped with a wireless interface in their computer. In late 2003 this situation calls for the use of 802.11b, since it is the most pervasive nomadic wireless standard available. This will probably not be the case in the future, as 802.11g, 802.11a, and 802.16 begin to permeate the marketplace. For the moment (2003 to 2004) a cheap consumer quality 802.11b AP may be the perfect solution for the hotspot environment because it is inexpensive and includes network functions such as a low-end router, DHCP server, and NAT functionality. Thus it is a one-box solution for this particular environment.

In a large office LAN environment, the usage characteristics may be quite different. In addition to Internet and email access, the system will probably be used for transferring files and inter-network communications. In other words, the usage characteristics will be much higher; more like the usage characteristics of a wired LAN. Depending on the density of the users and the users' need for access to other systems (like a hotspot in a hotel or airport), 802.11a or 802.11g, with their greater throughput, may be better alternatives for serving this situation.

The WISP, on the other hand, is probably not best served by a traditional AP, or for that matter, 802.11. The customers will be geographically dispersed, leading to a requirement for large area coverage from a minimum of locations. The user will have usage expectations of this system that are similar to the expectations users have of wired equivalent services, like cable modem and DSL. The need for large area coverage means that the equipment will need to be tailored to high EIRP with high gain antennas, and an access sharing methodology much more robust than the CSMA/CA scheme associated with the 802.11 standard. As of the time this book is being written, 802.16 equipment does not yet exist, so the choices today are either a high gain 802.11 variant, like Vivato or YDI, or one of the purpose built systems from the likes of Motorola, Alvarion, Flarion, Proxim, or Navini. I anticipate that there will be continued development of new equipment to meet the needs of the WISP industry, and that we will see more manufacturers offer equipment tailored to this business.

Further expansion of the concept of matching equipment capabilities to the needs of the business and the user can be seen in Figure 5-3. This figure is a matrix that overlays typical user and system characteristics with equipment options. It can assist you in selecting the most appropriate technology to serve a particular user community.

Identification of Equipment Requirements

After gaining an understanding of the customer needs and expectations, the implementer should be able to determine what equipment meets those needs. As described above, the needs of the customer are a key driver of equipment selection, however they are not the only driver. Size and environmental requirements, cost, manageability, reliability and availability all enter into the equipment decision matrix.

In the hotspot system, equipment should be selected that is compatible with the equipment that the users will have previously installed in their computers. This would drive equipment selection toward equipment operating on the prevailing standard adopted by the user base. Additionally, the equipment should integrate most of the network features that will be necessary to interface the equipment to the world, and to manage customers as they connect to the system.

The large LAN system will have its equipment selection driven by a more complex set of issues. The standard selected will have to be a balance between the now prevailing technical solution that is ubiquitous and low cost, and whatever standard is currently emerging as the next generation solution. This emergent solution will probably have a higher cost, since it has not yet reached mass appeal,

Equipment Characteristics														
	Designed for Coverage		**CPE**					**Base Station Hardware**						
Subscriber Service Type	Indoor	Outdoor	Internal antenna	External antenna	Weather proof	AC power	Battery	External antenna	Weather proof	AC power	DC power	High EIRP	Standards-based	Proprietary
Fixed	N	Y	N	Y	Y	Y	N	Y	Y	Y	?	Y	Y	Y
Nomadic	?	?	Y	N	N	N	Y	Y	Y	Y	?	?	Y	N
Mobile	Y	Y	Y	N	N	N	Y	Y	Y	Y	Y	Y	Y	?
Hotspot	Y	N	Y	N	N	N	Y	?	N	Y	N	N	Y	N
Office LAN	Y	N	Y	N	N	Y	Y	?	N	Y	?	N	Y	N

Figure 5-3: User needs matrix

and may also have some developmental wrinkles that still need to be fixed. On the other hand, it will also have greater capacity, greater spectral efficiency, and may offer better coverage and greater flexibility in deployment. Depending on your forecast of future usage demand and the need to allow your users to "roam" to other public or private locations and use their wireless access, one technology will be an appropriate, if not optimal, selection.

In this large LAN network, the additional network capabilities that were important to a hotspot are not necessary. Since this equipment will connect to an existing network, it is probable that all the advanced network features will be performed elsewhere on the network, and that the wireless equipment will only need to provide the wireless interface, and provide excellent remote management and fault isolation capabilities.

The WISP system has even more complex needs. Since standards like 802.11 were designed for wireless LAN type networks, they have not been optimized for serving a large area with dispersed users. While 802.11 has been made to work in these systems, other systems that were designed for use in this type of environment may offer a better technical solution. The trade-off here is cost. Because 802.11 is a mass-market product, equipment is very inexpensive and commonly available. Though a proprietary solution may be technically a better solution, it may be far more costly than using the 802.11 standard equipment.

Additionally, because of the nature of a WISP operation, the equipment must be located outside. This means that equipment must withstand the rigors of weather and an outdoor environment. Commonly available hardware designed for home or office use is not capable of surviving in the outdoor environment, so any equipment used in this environment will have to be located in a protected enclosure or will have to be designed for outdoor use. This leads to a set of long term maintenance issues, especially if equipment is mounted on towers: if the active electronics are located atop the tower, then the equipment will need to be removed and taken to the ground for service. Obviously, this is an additional ongoing cost to be considered in deciding the appropriate equipment design as well as the best location for the equipment.

Ultimately, any equipment solution has been designed with a set of expectations in mind, and has its appropriate place in the market. You need to understand the requirements of your system and the design intent of the equipment you are considering. Figure 5-4 provides an example of a comparison matrix that can assist with the identification of a solution that closely matches your needs. By completing such an analysis, you can be assured of selecting the manufacturer and solution that is best suited to serving your unique needs.

	Operating Band		Suitability for serving environment						Costs	
	Licensed	Unlicensed	Office LAN	Hotspot	Fixed WISP	WISP	Mobile Data		Base Station	CPE
802.11b Traditional AP		Y	○	○	◡	●	●		low	low
802.11b High Power PtP		Y	●	●	◡	◡	●		high	medium
802.11g		Y	○	○	◡	●	●		medium	medium
802.11a		Y	○	○	◡	●	●		medium	medium
802.16	Y	Y	●	●	○	●	●		high	high
802.16e	Y		●	●	◡	○	○		high	high
802.20	Y		●	●	●	◡	○		high	high
CDMA2000	Y		●	●	●	◡	○		high	high
Proprietary Solutions	Y	Y	●	●	○	○	?		high	high

LEGEND

○ optimal

◡ suboptimal, but useful

● not optimal

Figure 5-4

Identification of Equipment Locations

Now that you've narrowed the field of equipment options to a few that appear suitable for your system, we can begin to look at how and where to deploy sites to achieve the coverage and capacity desired in the system. Determining optimal antenna locations is the key to a successful deployment. An optimal location serves a multitude of needs: it provides optimal RF coverage; meaning it can be optimized to provide sufficient coverage of the area without leading to significant interference elsewhere in the system, it has easy access to power, it has easy access to network interconnect facilities, it can be easily installed and secured, and it has reasonable access for future service needs.

With the equipment selected, you have a baseline for the RF transmit power, receive sensitivity, and antenna options. These characteristics are used in conjunction with predictive modeling tools or survey tools to determine the area that can be covered with the selected equipment.

The first order of business is to evaluate the previously identified available locations for their suitability in providing coverage to the desired area, and so decide which tool is best suited to your need and determine the coverage potential of your sites.

By its definition the hotspot is a small open coverage area, so the simple path loss calculation discussed in Chapter 3 is useable to determine if the system can cover the required area. In fact, in a hotspot system, finding a convenient mounting spot with available power and network connection is probably more important than finding an optimal RF location. Of course, just because I made this statement, your first attempt to build a hotspot will surely be in some unique and bizarre location where there are multiple unknown impediments to RF coverage. Even though the hotspot appears to be a simple deployment, it's still worth spending a little time validating your assumptions with a field survey.

Although the same estimation techniques used in a hotspot can be applied to a hotzone or LAN, determining optimal antenna locations becomes a little more complex. The size of the area to be covered, the user density within the space, and the layout of the space must be considered in order to optimize locations from a capacity and interference standpoint.

In the LAN environment, the primary usage of the system will be wired LAN replacement, thus the bandwidth requirements per user will be significantly higher than those of the casual user accessing the Internet. Depending on the user density in the covered area, you may find that a single base station may not have sufficient capacity to serve all the users in its coverage area. In this case it may be necessary to reduce power, relocate the base station, or change the antenna to provide a different coverage area that includes fewer users.

The first objective in designing such a system is to calculate the per user bandwidth requirements. Because there are so many opinions about how to accomplish this, I will not discuss it here. Use whatever method of calculation you are familiar and comfortable with. Once you know the average usage per user, you can determine the total users per base station by this simple calculation: I_t / BW_{user}, where I_t is the radio information throughput. Do not confuse this with raw device bandwidth. We need to use the achievable device throughput based on the types of traffic on the network. For example in 802.11b the channel is advertised to have 11 Mbps throughput. In reality this is the total channel throughput including all overhead. The information bandwidth of the channel is significantly less, more on the order of 4.5 to 6 Mbps depending on the types of traffic on the network. Also, because the 802.11b channel is a TDD channel, the total throughput is shared by both upstream and downstream traffic, meaning that another derating factor must be applied based on the mix of upstream and downstream capacity requirements. All this means that the real data throughput of an 802.11b system may be as low as 2 Mbps for bidirectional symmetrical usage.

BW_{user} is the average bandwidth requirement. This is not the peak requirement of the user, but either the average usage, or the predetermined lowest bandwidth level available to any user during peak usage periods.

For example, in a system with symmetrical uplink and downlink requirements, and an average bandwidth per user requirement of 200 Kbps, an 802.11 system will support only 10 to 15 users per AP (2 to 3 Mbps information throughput / 200 Kbps per user). Remember, 802.11 has 4.5–6 MBPS total throughout. Since this example uses symmetrical traffic, the 4.5–6 MBPS is shared by the uplink and downlink traffic, thus leading to 2–3 MBPS available in either direction.

Assuming a normal office environment with cubicles and walkways, the average space allocation per employee is 250 square feet. In this environment an 802.11b AP will cover about a 50-foot radius, or about 8000 square feet of area with maximum bandwidth. This area may contain up to 32 users. In this situation the coverage defined area exceeds the capacity defined area.

There are several solutions for this. Antenna selection and power reduction can reduce the area covered to one that is more in line with the capacity needs of the users. Alternately, this may be a situation where 802.11b is not an optimal technology selection. 802.11a or 802.11g may be more suitable protocols because of their higher throughput. Either 802.11g or 802.11a will provide about 5 times more bandwidth than 802.11b, however the area covered by this bandwidth will be smaller than the 802.11b maximum throughput coverage area. In this same environment, 802.11g will probably cover less than 3000 square feet with maximum bandwidth signal levels, while 802.11a may serve only 1000 square feet at maximum bandwidth due to the additional propagation losses associated with its higher operating frequency. Do remember that these technologies rate adapt to lower throughput speeds as the signal strength drops. It is entirely possible that the needs of users can be met over a coverage area associated with one of more of the data sub-rates.

This is a case where future growth and changes in usage characteristics should be considered as part of the selection criteria. If capacity needs are expected to increase over time, then a higher bandwidth standard like 802.11a or 802.11g should be considered instead of 802.11b.

Now that the capacity versus coverage issues have been addressed, you know how much area should be covered by each radio base station location. Now you can begin to plan the location of hardware to meet the needs of the deployment. Use the drawings, blueprints and other reference data you've collected in combination with a physical site review to identify locations where the equipment could be placed giving easy access to power and interconnect, easy access for maintenance, avoidance of utility walls and massive metal objects, as well as located centrally to the desired coverage area of each radio base station. Depending on the physical layout of the space to be covered and the availability of power and interconnect, a number of location options are viable: you could use an omni antenna located on the ceiling in a central location, or you could use a directional antenna located high up in a corner or along an outside wall and pointing toward the space to be covered. You might try drawing shapes consistent with the antenna pattern (circles for omni antennas, cardioids or teardrops for directional antennas) and scaled to an appropriate size to represent the desired coverage on the blueprints. Lay out the shapes on the blueprints, arranging them so as to provide coverage to all the desired locations. Then physically check the locations to assure that power and interconnect are easily available at the locations. If not, see where you need to move it to gain easy access to power and interconnect. As you move the locations to ease deployment, try to keep the spacing between the base stations as even as possible. This will make your frequency planning easier. You may also want to

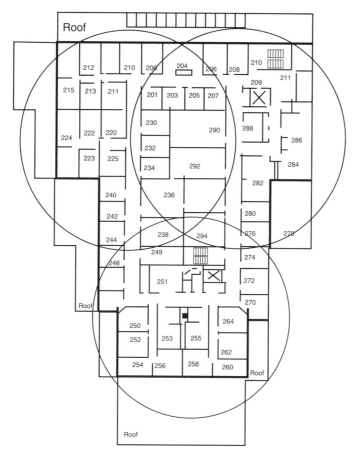

Figure 5-5

do a site survey on the selected locations. The reason is twofold: you can assure you get the expected coverage and, more importantly, you can determine the maximum coverage and therefore the interference area of each location. Knowing this will be useful when allocating channels to each location. Figure 5-5 shows one possibility for locating equipment to serve an office space. Depending on the unique needs and limitations of the space you are working with, such a solution may or may not be feasible for your deployment.

The WISP solution has much in common with the LAN system deployment. Coverage and capacity are both critical issues, multiple locations may be required to address coverage or capacity issues, and technology options will need to be considered according to system and cost requirements. The WISP system is designed to cover an extended area of potentially multiple square miles with a complex mix of terrain, morphology, and user locations. The system will be designed to provide Internet access service to residential and/or businesses within the coverage area. In addition, the location of the CPE is a consideration in determining the coverage of each site. If the CPE is located inside a structure, additional losses will be incurred, leading to a smaller reliable coverage area. Alternately, if the CPE can be located outdoors above the roofline, coverage distances will improve substantially.

Because of the size and complexity of the covered area, there is a very real possibility that users will be shielded from one another. This gives rise to problems using equipment that utilizes CSMA/CA, like 802.11-based hardware. Because CSMA/CA works by monitoring the channel prior to transmitting, the algorithm assumes that all users are close enough to each other to be able to hear each other. In the widely distributed WISP system, this is not true. This leads to a problem called the *hidden transmitter problem*, where users all have a clear path to communicate with the base station, but cannot hear each other. Since they cannot hear each other, they all assume the channel is clear, and try to transmit. The effect of this is that multiple stations to try to access the channel simultaneously, because they all think it's clear. In reality, the base station hears all the simultaneous users and cannot discriminate one from another. In essence the simultaneous users are all interfering with each other.

This behavior can be tamed using a feature in the 802.11 standard called *RTS/ CTS*, or Request to send/Clear to send. This is simply an additional protocol where a user asks the base station for permission to transmit, and waits until it receives permission before beginning its transmission. There are three downsides to this scheme. First, RTS/CTS is not a perfect solution. Collisions can still happen. Multiple stations may be asking permission near simultaneously, then all transmitting based on hearing a CTS belonging to a different user. Second, RTS/CTS takes overhead, because each transaction has to be preceded by a request and acknowledgment, thus leading to a further decrease in real channel data throughput. Third, since closer users have a stronger signal, it is quite common for those close in users to get an unfair share of the available bandwidth. This is because the closer-in users tend to have more signal strength at the base station, and the additional signal is easier to hear than the weak signal from the further-out user. In some cases it is possible for the close-in station to override the weak signal completely, and be the only signal heard by the base station.

Conversly, 802.11 equipment is cheap and abundant, so from a cost standpoint it may be the best solution for a particular need. Just be aware that as with any system offering service for hire, a WISP needs to assure that the users have fair and equitable access to the service. Clearly, even though the equipment is inexpensive, the issues surrounding the access methodology need to be given considerable thought. While it may be coerced into working, it is entirely possible that the solution will never be as good as a purpose built solution.

There are other systems that have taken the unique needs of large area access into consideration. 802.16 and 802.20 were designed with the needs of wide area coverage and possible mobile access in mind. So have some of the proprietary system standards that have been created by equipment manufacturers. These systems have the benefit of being designed for the express purpose of providing wide area access to multiple geographically diverse users. The downside is cost and/or availability. At the time this book is being written (Sept 2003), there is not yet commercially available 802.16 or 802.20 equipment, and the proprietary equipment that is available costs 5 to 50 times more than 802.11b equipment.

The wireless data field is still growing. New standards and vendors continue to evolve. I suggest that you carefully review the technologies and vendors available to you at the time you are designing the system. Determine their suitability to your unique business plan, and make your selection based upon your unique mix of cost, capacity, coverage, and spectrum availability requirements.

Once a technology has been selected, you need to review the coverage and capacity needs of your system. Because you are covering a large outdoor environment, propagation modeling can be an effective way of estimating coverage area, and assessing the coverage available from specific sites. Since the users cannot be identified by desktop location as they could in the in building LAN system, another method of estimating user density is needed. This is where demographics data can come into play. Such data is available from many sources, and has varying resolution. You can find demographics as coarse as an entire county, or as fine as fractions of a square mile. Use this data to determine how many households and businesses are in the geographic area associated with coverage. Then by using your expected market penetration percentage, you can determine the number of users in the area. You should already have an idea of the expected per user traffic, so you can now use the same formula we used in the LAN example to determine the traffic capacity needed for the area. Once again, determine if the covered area has sufficient capacity to meet user demand. If not, reduce the coverage area and add more radio sites to the system as needed to serve the territory.

Once you have determined your site locations, one other consideration you will face is connecting the sites together or to the Internet. Since the sites will have significant distance between them, CAT5 cable, which can only support 300 feet of link distance, is not a viable methodology to connect them together. There are several alternatives for interconnection to the Internet: purchase an individual access facility from a telco, CLEC (Competitive Local Exchange Carrier), or other provider for each of your radio sites, or use point-to-point radio to connect your facilities together and bring all traffic to a common location for delivery to the Internet.

The former solution may lead to a more robust system, because a single facility outage isolated only one site and its associated coverage area. It will also be more expensive because you cannot aggregate traffic from multiple sites in order to most effectively use the capacity of the facility you're paying for. Additionally, there will be more equipment necessary because each site will have to have it's own network hardware to provide access control, DHCP, maintenance access, and other network and security functions.

The latter solution allows all the network equipment to reside at a single location, thus reducing the need for redundant equipment at each site. It also leads to the additional cost of a facility to connect the sites back to the designated central location. This could be accomplished with a leased Telco facility, although you'll probably be limited to the T-1 or E-1 facility speeds of 1.544 or 2.048 Mbps, or multiples of this. Given that this may be a fraction of the bandwidth of the radio, it may not be an appropriate choice. A better choice may be point-to-point radio facilities. If the sites in question have clear line of sight to each other or to a central location, this becomes a feasible implementation. There is equipment available in both the licensed and unlicensed bands that can be used to provide this type of connection. Better yet, these radios can be had with an Ethernet output, so they can simply be plugged into your other network equipment without the need for specialized equipment to convert from Ethernet to some other standard like T-1 or E-1.

As discussed each of these solutions has trade-offs. Use the flowchart in Figure 5-6 as a basis to assist in selection of the most appropriate solution based upon your unique environment and needs.

Channel Allocation, Signal-to-Interference, and Reuse Planning

The number of available channels in a system will be predicated on three things: the spectral allocation available to you, the spectral requirements of the equipment you've selected, and other users or interferers in the band.

When you did the site surveys at your locations, one of the things you noted was the noise floor, and any other users in the band that you noted. If possible, avoid channels with existing users or a high noise floor.

In the hotspot system this is simple: pick the quietest channel and implement it as the operating channel of your equipment. Since there is only one radio base station or AP, you're done.

In the office LAN or hotzone, you must consider not only outside interference but also the interference you will self generate when all the sites are active. This means that you must carefully allocate channels in order to minimize interference between locations. In general, get as much physical separation as possible between co-channel locations. As detailed in

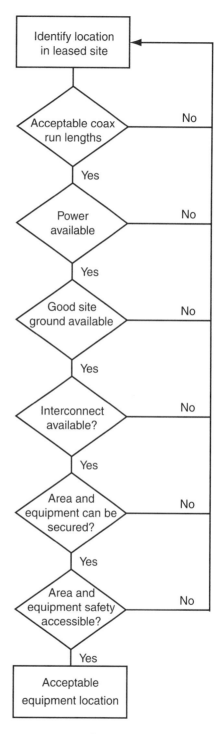

Figure 5-6

151

Chapter 4, the building layout and coverage information collected during the site survey will be invaluable in determining how to allocate the channels based on the actual coverage and overlap of each location. Ideal channel separation, i.e., sufficient so as to cause no interference, is rarely achievable. The best separation that can be designed for is separation sufficient to provide adequate C/I margin to the majority of the coverage area of each base station. The C/I requirements of the system will vary based upon the technology. If the manufacturer does not publish C/I requirements they can generally be thought of as identical to the S/N requirements of the technology, which by the way are normally the published receive sensitivity numbers. The receive sensitivity is nothing more than the amount of signal required for the equipment to perform at a certain threshold level. The reference against which this is measured is thermal noise in the channel.

What does this mean to the design? It means that the coverage and/or capacity performance of the network will not be as good as in would have been in an interference free environment. The required signal strength will increase by the number of dB the interference has raised the noise floor.

For example, the noise floor of a 20-MHz 802.11b channel in the 2.4-MHz band is −100.43 dBm as calculated by the thermal noise equation. The published receive sensitivity specifications are based on this noise floor. If interference adds undesired signal (man-made noise) to the coverage area, the noise floor increases above the level contributed by thermal noise alone. The basic receive sensitivity does not change, but the system performance does. For every dB interference adds to the noise floor, the perceived receive sensitivity will be worsened by an equivalent amount. In the case of an 802.11b device with a published 1 Mbps sensitivity of −92 dBm, this sensitivity is based on an expected noise floor of −100.43 dBm. If interference raised the noise floor to −98.43 dBm, the device would no longer perform when the signal strength was −92 dBm. The interference has raised the noise floor by 2 dB, so the new signal requirement would be −90 dBm for the device to offer the same 1 Mbps performance level.

This is another reason why the site survey is a useful system-planning tool. By knowing the signal level contributed by other locations, you can assess how much interference adds to the noise floor. This allows you to estimate the real coverage of locations based on the additional noise level caused by co-channel users in the form of interference.

The WISP system requires the same diligence in allocating channels for the same reasons as those in the LAN environment. Interference, regardless of its source, will negatively impact either the coverage potential of a site or its ability to provide maximum throughput over its designated coverage area. Pick the channel with the lowest noise floor, and use antenna aperture, downtilt, and site separation distance to assure sufficient isolation of co-channel signals.

Antenna downtilt is another subject that the WISP operator should become familiar with. Because the WISP system is located outside, and is probably located at some elevation above ground, downtilt becomes an important factor in optimizing coverage and minimizing interference. Think about it this way: the antenna you will use has a vertical polarization, and has a main lobe with the highest energy density pointed perpendicular to the antenna orientation. This means that the main beam will be pointed horizontally, 90 degrees shifted from the mounting orientation of the antenna, in other words at the horizon.

As you can see in Figure 5-7, the main beam points at the horizon, and it is entirely possible to "miss" the intended service area with the main beam and to end up

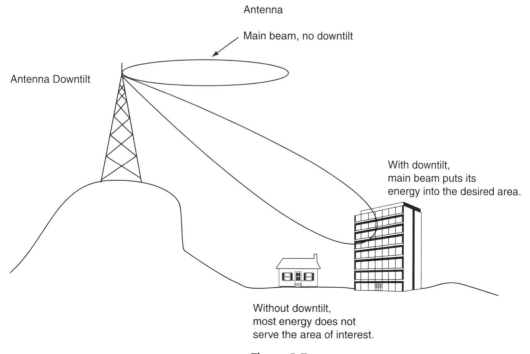

Figure 5-7

serving the desired area with side lobe and sub lobe energy. This situation leads to impaired service in both the desired coverage area and surrounding areas, because the main lobe energy misses the target user, and increases interference in adjacent areas because the main lobe energy is pointed toward the horizon, and other potential users.

Downtilt corrects both of these situations by pointing the antenna's main lobe toward the desired coverage area. This can be accomplished two ways: mechanically and electrically. Mechanical downtilt is used only with directional antennas, and is accomplished by physically mounting the antenna in such a way as to tilt it towards the ground by some number of degrees. Electrical downtilt is achieved by the design of the antenna, and can be applied to both directional and omni directional antennas. In fact, electrical downtilt is the only way to implement downtilt in an omni antenna. The number of degrees of downtilt is calculated with the simple formula: **arctangent (H/D)** where H is the effective height and D is the distance to the far edge of the desired coverage area.

To keep the calculation simple, the formula is based on calculating the value of the adjacent angle of a right triangle, therefore the effective height is determined as the difference in elevation between the antenna and the area to be covered, and the distance is the physical distance between the antenna and the far edge of coverage in the same units used for height measurement. For example, let's take a case where the antenna is mounted on a tower, which is on a hill overlooking the coverage area. The ground elevation at the edge of the coverage area is 540 feet, the top of the hill is 650 feet, and the antenna is mounted at the 100-foot level on the tower. In this case, the effective antenna height is (650 + 100) − 540, or 210 feet. If the distance to the far edge of coverage is ½ mile, then **atan(210/2640) = 4.548 degrees of downtilt**.

As shown in Figure 5-8, by down tilting the antenna by 4.5 degrees, the center of the main beam is aimed at the users furthest from the site, thus maximizing the energy density in the area furthest from the site. Areas closer to the site have less path loss due to distance, and are effectively served by the lessening energy density of the main lobe and sub lobes of the antenna. The other major benefit of downtilt is interference reduction. Without downtilt, the desired coverage area had less than ideal signal, while undesired areas were getting most of the site's energy.

By using downtilt to maximize energy density in the desired coverage area, a side benefit is that this energy no longer gets to places it does not belong, and therefore does not appear as interference in some undesired area.

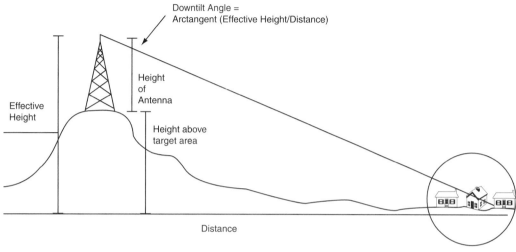

Figure 5-8

Network Interconnect and Point-to-Point Radio Solutions

In small venues like a hotspot or small office, network connectivity should not be a significant issue. CAT 5 cable is an inexpensive and readily available solution for connectivity in these small area locations. Even large office spaces can have the radio solution effectively networked using CAT 5 cable. The distance limit for CAT 5 cable is 300 feet per run, which allows quite a large area to be served by cable alone, with no need for intermediate regenerators. If greater distances are needed, it is feasible to divide the network on a floor by floor basis, and run cables to a central point where they are connected using a switch or router, which is in turn connected to the wired network in the building. The decision of how to cable will be dependent on the existing wiring, existing network, and construction of the property.

But what about the campus environment, where the space to be served with wireless is in a number of disparate buildings, or the WISP system that covers a community from a number of different sites? In these cases, the distance limits associated with CAT 5 make it unusable. In the campus environment, Ethernet over fiber (10Base-FL and 100Base-FX) connections may be feasible, depending on the availability and cost of duct space between buildings. In the WISP system where such ducts between sites are rarely available, or in the distributed building office environment where no ducts are available, another interconnect option needs to be considered. That option is of course radio. Not only can radio-based systems be used for connecting users, they can be used to connect together the disparate locations.

Radio facilities such as these are called *point-to-point links*, and can use a variety of licensed and unlicensed bands for operation. Unlicensed band links commonly offer limited bandwidth (1 to 50 Mbps), while licensed microwave bands can offer links with hundreds of Mbps throughput. When a single central facility is used as a distribution point for a number of remote locations the resulting network is known as a "hub and spoke" configuration. Figure 5-9 illustrates such a network.

These links can be designed with the same tools you use for designing the radio networks for the users. The big difference is that there are only two points to connect, they are both known points that do not move, and they should be within line of sight of each other. This eases the design because you need only consider the coverage at two points in space that can see each other. If the locations have clear Fresnel zones, Free Space losses can be applied; if the Fresnel zone is cluttered then line of sight loss characteristics should be used. Because there are only two locations, highly directional antennas can be used with the radios. This helps overcome the path loss as well as reduce unwanted interference by keeping the energy tightly focused on the other station. In fact, these antennas can have less than two degrees of aperture, depending on their size and frequency of operation. These small apertures mean that mounts must be sturdy, and the antennas must be carefully aimed at each other to assure the stations are initially, and remain, accurately pointed at each other.

A new concept to consider on these links is redundancy. Because these links are used to connect traffic-bearing locations together, a failure of one of these links will isolate those traffic-bearing locations from the rest of the network, thus leading to user communication failures. This single point failure can be remedied by making the interconnect links redundant. This can be accomplished in several ways.

The first method is the simplest but perhaps most expensive way to accomplish redundancy: use redundant radio equipment. This can be done in two ways, the first

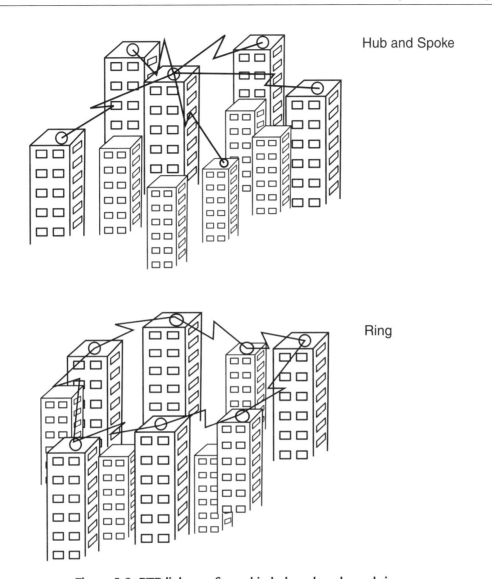

Hub and Spoke

Ring

Figure 5-9: PTP links configured in hub and spoke and ring

being to have two active independent radios each bearing half the total traffic. The advantage is that a failure will not isolate the end site, but may, depending on the total throughput capabilities of the radios, reduce the total traffic capacity that can be handled by the remote site. The disadvantage is that two independent channels are needed, and equipment costs are significantly increased due to the need for two radios. Another redundancy method is "Hot Standby." In this method, there is still the need for two independent radios, but only one is active at any given time.

If the equipment senses a fault, it automatically switches to the standby radio and allows traffic to flow unimpeded. There are two advantages to Hot Standby: only one RF channel is needed, and a failure does not impact traffic flow from the remote site. The disadvantage of cost increase over a single radio still exists in this scenario.

Full Redundancy and Hot Standby redundancy are commonly used when only a single remote site is connected, or there are multiple remote sites that cannot be connected to any other location. If more than one remote site needs to be connected and there are multiple choices for paths to connect them, another network topology called a ring should be considered.

In the ring network a single site connects to remote site one, remote site one connects to remote site two, and so on. The last remote site connects back to the origin site as shown in Figure 5-9. In this scenario, traffic can flow around the ring in either direction. A single failure cannot isolate a site, because there is an alternate path to route the traffic. There are several disadvantages to this topology: Backhaul facilities need to be large enough to handle the traffic presented by multiple sites, additional network hardware must be configured in each site in order to drop and insert traffic from the site, and to manage the traffic flow on the ring, and finally, the locations need to be arranged in such a manner as to accommodate the ring topology. The primary advantage of the ring architecture is cost: one extra radio per ring is needed to achieve redundancy, instead of one extra radio per remote site. Another advantage of a ring architecture is that it can be used to extend the service area far beyond the distance limits of a single radio link. Considering each link as an independent starting point allows you to develop a network that continues to expand outward from each site on the ring. The only limitation is that at some point the network needs to loop back in order to close the ring.

From a frequency utilization standpoint, it is best to use different frequencies for the backhaul and network connections than you are using for connecting users to the system. Once again this has to do with potential interference generated by the additional facilities using common channels. Even though the backhaul facilities use directional antennas, there is still a possibility for mutual interference. In fact having backhaul in a completely different band sometimes makes sense because it allows all the channels available in your primary band to be used for providing service to users, thus allowing growth of capacity in the network. Also there may be other bands more suited to the needs of back haul because they have the ability to connect over greater distances or provide greater capacity.

Costs

The last discussion of this chapter might actually be your first consideration: what's it all going to cost, or looked at another way, how much can I afford to spend and what compromises will be necessary? Costs are largely broken into two categories CAPEX and OPEX. The CAPEX costs are one time costs associated with capital equipment, construction, design and planning, and so forth. OPEX are recurring costs associated with such things as interconnect, leases, utilities, maintenance, and other month-to-month costs.

Network design will have an impact on all these cost elements, and the final network design will have to include the trade-offs associated with costs as well as performance. As I've said before, there is no free lunch. For example, using cheap equipment to reduce CAPEX may have a serious negative impact on OPEX, because equipment reliability or maintainability suffers. There can be many consequences of CAPEX decisions that in the end have a significantly greater cost effect on OPEX. Be careful when making CAPEX financial decisions and make sure you consider the long-term ownership costs too. You may have to live with the aftermath of your capital decisions for a very long time.

The Five C's of System Planning

This and previous chapters have discussed elements of system selection, design and performance. They have also mentioned the fact that trade-offs are necessary when selecting a solution or planning a system. All of those discussions have led us to here: the five fundamental aspects of a real-world communication system. In its simplest form, there are five elements that will affect your decision on what technology to select and implement. Those fundamental elements are Cost, Coverage, Capacity, Complexity, and C/I ratio. Cost includes both initial and ongoing costs of ownership, coverage can be either the total area to be covered or the area covered by a single base station, capacity can be either the necessary capacity or the capacity of an individual channel or base station, complexity can be defined as the overall size of the network (for example how many individual sites and discrete pieces of hardware are necessary to make the system function), and finally C/I or the amount of interference that needs to be tolerated. This interference can come from two sources: it can be internally generated through channel reuse in the system, or it can be generated by other systems over which you have no control.

These elements are inextricably interrelated and are almost mutually exclusive. For example, you cannot have a cheap but complex system, nor can you have a simple and inexpensive system that also has great coverage and capacity. The selection of a suitable system solution will be driven by an understanding of the business needs, the user expectations, and area to be served then balancing those needs against a prioritization of the five C's. This balance will be different in each situation. Always remember that a successful set of trade-offs in one situation may lead to a dismal failure in another situation.

Use these five elements when initially formulating your system requirements. Ranking these factors in order of importance to your business helps to organize your selection process by allowing you quickly to eliminate choices that obviously do not fit your hierarchy of need. For example, if low cost and maximum coverage are the most important criteria, a single site high power solution might be the best solution for your need, providing equipment and spectrum for such a system could be obtained. On the other hand, if high capacity and dealing with a limited coverage area and hostile interference environment are most important, then you may need to deal with the additional costs and complexity of a network requiring multiple sites in the coverage area. This of course gives rise to considering the number of channels needed to support the multiple sites and an analysis of whether the reuse required by such a plan can be managed.

As you see, the questions and considerations quickly multiply, and each time you make one decision, you may eliminate or severely modify another of your assumptions or requirements. Learning the relationships between the five C's as driven by your unique system requirements will allow you to make informed decisions and optimize the necessary trade-offs into a system that best suits the financial, user, and operational needs of your particular situation.

CHAPTER 6

System Implementation, Testing and Optimization

- Real-World Design Examples
- Example One: Local Coffee House
- Example Two: Office LAN Deployment
- 2.4 GHz RF Coverage Results
- 5.6 GHz RF Coverage Results
- Capacity Requirements
- System Design Analysis
- NEC, Fire, and Safety Code Concerns
- Example Three: Community WISP
- Community: a Garden Style Apartment Complex
- RF Considerations
- Backhaul
- Weatherproofing
- Grounding and Lightning Protection
- Community: A Small Area Subdivision
- Equipment Selection
- System Planning
- Community: An Urban or Suburban Area Serving Business Users
- Spectrum Issues
- Design Considerations
- Community: A Small Town System for Consumer and Business Users
- Example Four: Mobile Broadband Network
- Initial Modeling
- Preliminary Information
- Coverage Modeling
- Capacity Modeling
- Cost Modeling
- Designing in the Real World

System Implementation, Testing and Optimization

Real-World Design Examples

Now that the basics have been covered, let's take a look at putting the concepts into practice by designing real systems. In the following pages I'll take a real example of each of the system types we've discussed and walk through the decision process that resulted in a deployed network. Additionally, I'll provide a look at the design issues and approaches necessary to deploy a fully mobile network.

Example One: Local Coffee House

The owner of this venue wanted to expand the available services in the shop to include charge-free Internet access for patrons. Internet access will be provided by one shop-owned hardwired computer, plus wireless access for patrons with laptops or PDAs. The area to be covered is a space 25 feet by 40 feet with 15-foot tall ceilings. The area is open with no obstacles except for the service counter located along one side. Initial and recurring costs need to be as low as possible, since this is an overhead service, not a revenue generator.

Walking through the design process yields the following issues and answers: The coffee shop wants to own and deploy the wireless solution, so a Part 15 compliant system is necessary. Further it must be a low cost solution, so 802.11b, 802.11a, and 802.11g are viable solutions. Given that it is expected to work with user customer's laptops and PDAs, the selection is further narrowed to 802.11b, which at the time of implementation is the most widely available and commonly used standard. The area to be covered is relatively small and open, and the bandwidth requirement of the customers is minimal, so a single 802.11 AP should provide adequate coverage regardless of its location in the room.

So this quick review of the basic needs has already determined the most appropriate RF technology: 802.11b. Now let's look at the other needs. There is no existing network, so one will have to be built in order to support the combined wireless and wired solutions. Since the service is offered to the patrons at no charge, there is no need to secure the network or to provide RADIUS service for user name and password authorization. This network will need to provide DHCP service to support the nomadic wireless users, and should also provide Network Address Translation (NAT) service in order to eliminate the need for numerous public IP addresses. It also needs to provide both wired and wireless support.

The network requirements can be met with a simple 4-port router, a function that can be purchased as a stand-alone device. Alternately, this function is also built into many inexpensive consumer grade 802.11b APs. In fact, the integrated router/ AP solution was selected for this deployment due to its lower cost, reduced network complexity, and reduced maintenance. The selected device, a D-Link DI614, is a combination 802.11b AP with built in 4-port router, DHCP server, NAT capability, and a host of other screening, filtering and security capabilities that were unnecessary for this application. Best of all, it was sourced from a local computer store for less than $90.

The network will be used for casual web browsing and email, and it is expected that the maximum number of simultaneous users will be five, so the traffic load will be quite low. Internet connectivity can come from either the local CATV company in the form of a cable modem or from the local RBOC in the form of a DSL Line. Either will deliver a single IP address, connection speed of at least 256 Kbps, and Ethernet connectivity. In this case, the cable modem-based service was selected due to its lower cost.

The location of the hardware was ultimately dictated by the location of a CATV drop in the building. It just so happened that the drop was in a back corner very close to where the wired computer was to be located. Netstumbler was used to check for interference and to verify coverage with the AP setting on a table in this location. The result of the interference test showed that there was another AP working in the vicinity, and that this AP was tuned to Channel 6. In order to avoid interference, channel 11 was selected for and programmed into our AP. The resulting Netstumbler coverage survey showed excellent coverage at all tables in the shop. Since the location was known to work at a three-foot elevation, it was expected that a bit of additional height would not hurt the coverage, so in order to keep the equipment out of reach of curious hands, a shelf was placed on the wall at a ten-foot elevation, the cables were routed behind a piece of wood trim molding in order to hide them from sight, and the Router/AP was placed on the shelf.

This solution met the needs of the client extremely well. It was low cost, low maintenance, easily installed, and is working well for the regular patrons who stop by with their Wi-Fi enabled computers.

Example Two: Office LAN Deployment

This is a significantly more complex scenario as compared to the coffee shop example. In this case a company is expanding, and is looking at ways to supply network connectivity in the new space, as well as methods to provide access to their company network in the new space. The new space is physically separated from the company's original location by approximately ½ mile. The space to be covered encompasses 40,000 square feet on multiple floors of a building. There will be about 200 computer users in the new space, as well as employees visiting from the main space. The employees use the network for email, some Internet access, and access to central databases.

The first step in analyzing the needs of this network is to better understand the needs and expectations of the users, the IT department, and the budget associated with the project. From this information collection process it was determined that there was line of sight path availability between the two buildings, so a point-to-point facility could be used to connect them together.

It was also discovered that the majority of the computer users in the company used traditional desktop computers, and that there was existing CAT5 wiring in the new space. Another issue uncovered was that the management users, who were the only employees with portable computers, were currently hardwired to a CAT5-based network, and were hoping to get full portability in their old space as well as the new space, and additionally wanted to be able to take advantage of the wireless facilities that were becoming more common in hotels, airports and other locations where they frequently traveled.

The IT department did not want to fully adopt a wireless solution. They felt that CAT5 was a fine, inexpensive, secure alternative for the majority of the users, since these users were desktop computer users and the cabling existed anyway. They were also concerned about the security of wireless solutions. Finally, they were also planning to move some server hardware to the new space in order to provide physical separation of equipment as part of their disaster recovery plan. This meant that the network connection between the buildings would need to be extremely stable, secure, and provide at least duplex 100 Mbps throughput.

Because of the complexity and number of unique issues to each subcomponent of the project, it was reasonable to break the project down into three subcomponents: the inter-building link, the user connectivity in the new space, and a wireless overlay deployment in the original space.

Given the input on usage, the inter-building link has to be high-speed, reliable and secure. This link could be provided using a wireless connection, since a line of sight path existed between the rooftops. Another alternative is using a facility leased from the local Telco.

The Telco option was explored first. The telco could not provide a 100Base-T Ethernet solution. They could only provide standard telco facilities of T1, T3 and OC3. Because of the speeds required, multiple T3 facilities or an OC3 facility were the only solutions. Neither existed, but the Telco determined that the capacity could be made available in under six months. The cost of the OC3 facility would exceed $5000 per month. Additional costs would be incurred for the routers that would be needed on each end to convert between the computer network's 100Base-T protocol and the telco's OC3 protocol.

The second analysis was on radio-based facilities to connect the buildings. Although unlicensed product options like 802.11a could have been used that would have directly interfaced to the 100Base-T computer network, and would have cost less than a month lease on the OC3, it was determined that these solutions were too limited in bandwidth and were not as secure as necessary. The other alternative was to use spectrum licensed under FCC Part 101 rules to build a licensed microwave link between the buildings. This would provide the speed and security desired, and could be designed to achieve over 99.999% uptime (less than 34 seconds per year of downtime). Unfortunately, it would also be a far more expensive alternative.

Several products were reviewed, and several were found that could provide 100Base-T connectivity, as well as providing some T1 connectivity. This had the additional benefit of being able to provide facilities for voice connections back to the main PBX, thereby avoiding another set of monthly telco charges. The downside of this solution was cost. In order to get a license for the facility, an engineering and coordination study would need to be accomplished, and a license application would need to be made to the FCC. Such a study and license application would cost $3,000 to $5,000, and the hardware would cost around $27,000. Given the timing of the coordination study and FCC filing, this solution was deliverable in about the same time as the telco solution, and would pay for itself in half a year, based on the monthly telco facility lease costs.

While a $30,000 solution seems extravagant when compared to a sub $2,500 unlicensed link, the unlicensed facility has some severe constraints. First, an 802.11a link has a maximum throughput of 54 Mbps total, while a duplex 100Base-T connection has 200 Mbps of total throughput (100 simultaneous each way). This means that to get comparable bandwidth from an 802.11a solution, four radio links would need to be simultaneously operated on each rooftop, and the network would need additional hardware to split and combine the traffic from the multiple radios. On the positive side, there are sufficient channels available to accomplish this, and such a multiradio solution has an additional benefit of providing redundancy.

So in reality the single $2,500 802.11 link turns into a $10,000 plus solution that uses significantly more rooftop real estate, and has no guarantees that its performance will remain at current levels as other users begin to migrate into the 5 GHz bands. Additionally, the security requirements desired by IT were too stringent for a stand-alone 802.11a product. In order to use 802.11a, additional dedicated encryption hardware would have been necessary on the links, thus adding additional expense to the solution.

The licensed solution was selected due to its ability to meet or exceed the security and bandwidth requirements of the network, assure a long term solution that would not be impacted by other users in the band, and offer the additional cost savings associated with carrying internal PBX traffic.

With the inter-building link out of the way, it's time to focus on connectivity in the new space. Since the majority of the computers on the network will be desktops, and CAT5 cable already exists, the IT department decided to use wired connectivity for the majority of the computer users in the space. The need for wireless connectivity would therefore be limited to supporting about 20% of the workforce who had portable computers and a need to "roam" between the buildings. This also means that the original office space would need a wireless overlay added in order to give these users equivalent connectivity in either location.

Based upon the desire of the users to have the ability to use "hotspots" in hotels and airports while traveling, the technology choice for the wireless application was limited to one that was 802.11b compatible. This proved to be a nonissue since several manufacturers are producing PCMCIA-based cards that incorporate operation on 802.11a/b/ and g standards and bands.

The network hardware required more diligence. The solution needed to be flexible in allowing external antennas and power control to be used to tailor coverage, and the solution needed to be remotely manageable so IT could monitor and maintain all the APs from their network monitoring center. There are a multitude of consumer quality 802.11x solutions available at very low cost, but these solutions lack the manageability and antenna/power flexibility that are needed in a commercial deployment. These factors limit hardware selection to enterprise class equipment.

Both the Proxim AP2000 and Cisco Aironet 1200 solutions offered external antenna connections, central management, and 802.11 a, b, and g support. Cisco hardware was selected in large part due to an existing relationship between the IT department and Cisco.

The next task is to determine which standard is most appropriate to covering the target area. Assessing the physical area showed the floor space to have a square shape with a central elevator core and walled offices and conference rooms around much of the outside. The open area is comprised of cubicles. There is 10,000 square feet of space per floor and four floors encompass the new space. Future plans call for expanding into more space in this building. To assess the coverage of the various standards, an AP supporting the various standards was used for the site survey.

Prior to testing, a spectrum analyzer was used to look for signals in the bands. Several signals that appeared to be wireless PBXs or cordless phones were found in the 2.4 GHz band. The wireless PBX system signals appeared to be frequency hopping systems with their energy concentrated in the lower portion of the band. Additionally, several existing 802.11b systems belonging to other tenants in the surrounding buildings were identified in operation. The AirMagnet analysis tool was used to gain more information about the other 802.11 systems. There were a total of 13 unique SSIDs found. Six systems were operating on channel 6, three on channel 1, four on channel 11, and one on channel 4. The measured RSSI of these systems varied from –87 dBm down to noise floor, with the strongest signals being very localized near certain windows and in certain offices. There were no signals noted in the 5 GHz band.

The coverage potential of APs functioning on 2.4 and 5.6 GHz was assessed, using several discrete locations and antennas. The first test was conducted using the AP connected to a 90 degree beamwidth antenna, which was mounted at ceiling height in an outside corner and aimed toward the center of the building. Coverage was measured on the floor. Additional measurements were made on the floors above and below and outside the building in order to assess the amount of signal leaking into undesired areas.

The second test was conducted with the AP connected to an omni antenna that was mounted on a ceiling tile in the center of one side of the open area between the elevator core and the offices. Again coverage was measured on the floor, on the floors above and below, and outside.

2.4 GHz RF Coverage Results

Initial measurements showed that an AP operating at 2.4 GHz (thus supporting either 802.11b or g) provided excellent coverage of the open areas and good penetration of the offices and conference rooms from either location. The corner mount with 90 degree antenna provided the best coverage area in the open area, and minimized the signal seen outside the building in the parking lot. Coverage in the offices was useable over about ½ the length of the building, but signal strength was significantly lower than the signal in the open area that reached all the way to the far wall. Coverage from the omni location was also good, although there were some weak spots in the open area produced by the shadows created by the elevator core. The coverage in the offices and conference rooms was better with the omni, mostly because the antenna was "looking" in the glass windows of these spaces, rather than looking at reflected signals or signals received by the "edge" of the antenna's main beam and attenuated through plasterboard interior walls, which was what occurred when using the 90 degree antenna. Unfortunately, the signal from the omni antenna could be seen in more places outside the building. This was expected, because the directional antenna was radiating toward the interior of the building, whereas the omni was radiating in all directions, including through the windows to the outside. Interestingly, this is the same reason coverage in the offices was better, so once again, a trade-off will need to be made: Is interior coverage with minimal equipment more important than signal leakage and its attendant security concerns?

Inter floor coverage was also analyzed from both locations. This was necessary in order to determine the interference signal levels that would be created in the other targeted coverage areas. Areas of unintentional coverage were measured from both locations on both the floor above and the floor below. The areas of unintentional coverage manifested themselves close to the windows near the location of the APs, which led to the assumption that the signals were "ducting" along the metal window frames, and radiating into the adjacent floors. The signals were strong enough to need frequency reuse coordination in order to limit system degradations due to interference.

5.6 GHz RF Coverage Results

The coverage potential of the same locations varied significantly when the APs were operated as 802.11a devices in the 5.6 GHz band. The achievable coverage dropped in both locations. The corner location provided almost no coverage in the offices, and coverage in the open space fell off by about half. Coverage outside was minimal. The centrally located omni fared much better. The open areas were covered well, and the offices were covered with sufficient signal to assure connectivity, although not at the full 54 Mbps rate potential of the 802.11a technology. Coverage outside was significantly (>15 dB) lower as compared to 2.4 GHz. Interference on adjacent floors was also reduced when using 5.6 GHz.

The behavior at 5 GHz accurately reflects the expected coverage change due to frequency increases. Remember, Figure 3-1 showed that if you double the frequency the free space path loss increases by 6 dB, and a 6 dB variation in signal strength would double or halve the coverage distance. That's the effect we're seeing here. 5 GHz RF experiences greater path loss as well as greater attenuation when penetrating objects.

Capacity Requirements

Since the majority of the network users will be desktop users, IT decided that those devices should be hardwired to the network, thus leaving wireless to connect only management, IT, and sales employees, who are the company's only laptop users. The total number of regular users in the new space is 50. Additionally, there is expected to be another 5 to 10 users from the primary facility that would need access to bandwidth while visiting the new space. The usage characteristic of these users is very light, requiring less than 500 Kbps average to satisfy their needs.

With this limited loading, and assuming an equal traffic distribution per floor, the wireless system only needs to provide about 6 Mbps of traffic capacity per floor.

System Design Analysis

Given the current 6 Mbps per floor capacity requirements of the network, any RF implementation that covers the desired area will be adequate also to serve the capacity demands placed upon the network.

From a technology standpoint, if there were no interference from either internal or external sources, 802.11b or 802.11g would require two APs with directional antennas mounted either in opposing corners or two APs with omni antennas centrally located on opposing sides of the building. The corner mounting minimizes the undesired coverage outside the building, but also reduces the signal in the offices and conference rooms, which are the very places the system will likely be used most. The central location provides better service to these areas, but also leaks significant signal out of the building.

Frequency reuse will be required since there are only three nonoverlapping channels available in the 2.4 GHz band, and the system will require a minimum of eight APs to cover the area. The self-generated interference from channel reuse plus the fact that there are numerous other signals from other sources, means that the minimal coverage solution of two APs per floor may not be the best solution because of the need for increased signal strength to overcome interference and allow all users, regardless of location, a full speed connection to the network.

Given the measured coverage, adding a single AP per floor will not solve the problem. Two extra APs would be required per floor. This is a significant expense. A lower cost alternative would be to use the dual radio capabilities of the selected product, centrally locate the hardware, and extend LMR400 cable to the antenna locations. With a central AP location, the cable runs would be approximately 50 feet each, thus adding 3 dB of loss. Selecting an antenna with 5 to 7 dB of gain easily overcomes the line loss and adds a modicum of extra performance to the system by improving link margins a few dB.

Making a reuse pattern with three channels, spread over four locations per floor, and multiple floors is challenging. Three channels are insufficient to design an interference free frequency reuse plan, although possibilities exist to use internal objects as shielding to provide interference protection. Channel 6 could be used on the east side, channel 1 on the west side, and channel 11 could be used on the north side and again on the south side with the expectation that the elevator core will provide sufficient shielding so as to avoid interference between the two co-channel APs. The next floor would have to shift the channel use because of the signal leakage between floors noted during the survey. The next floor could use channel 11 on the east side, channel 6 on the west side, and channel 1 on the north and south side. As you can see, no matter how you plan it there will be areas of interference floor to floor as well as potentially on the floor.

A better alternative might be to use a slightly overlapping channel plan with channels 1,4,7, and 11. This plan does have some channel overlap that will result in some energy from the adjacent channel "spilling over," thus resulting in some opportunity for degradation at very high traffic levels. However the spillover may have less effect than the co-channel interference will. In this design, channel 1 would be used on the north side, channel 4 on the east side, channel 7 on the south side, and channel 11 on the west side. The next floor would have the plan rotated 180 degrees, so channel 1 would be on the south side, 4 on the west, 11 on the north and 4 on the east. The next floor goes back to the first floor plan, and the next floor again gets the second floor plan. This plan provides maximum separation between reuse by allowing no reuse on a floor, and providing as much physical separation as possible between co-channel APs on different floors.

In order to serve the area with an 802.11a-based system, a four AP solution will be necessary. The coverage provided at 5.6 GHz was insufficient to allow a two location per floor network to perform as desired. Since the APs support two radios, a two AP solution with remotely located antennas could be done, but LMR400 will exhibit almost 6 dB of loss over the 50 feet of cable between the AP and antenna. Stepping up to higher quality, lower loss cable could be done, but the low loss cable is much larger in diameter, much stiffer and more expensive than LMR400.

Reuse planning at 5 GHz becomes easy because there are 12 non overlapping channels available, enough to have a fully independent channel implementation on three of the four floors, with reuse occurring only between the first and fourth floors. In addition, there were no other signals noted in the 5 GHz band during the survey, so managing and designing around outside interference is not an immediate issue.

So, we now have ten viable design alternatives that all should work. Each has different costs, capacities, implementation, and operations considerations. Choosing the optimal alternative depends on knowing the unique requirements of the user group and the financial considerations associated with the project. The best solution will be the one that comes closest to cost effectively serving the needs of the user while eliminating as many shortcomings as possible.

The first decision is to select the best option for the wireless network. Since security is a concern of the IT group that overrides cost considerations, the 802.11 b and g solutions using omni antennas are deemed unacceptable. The 802.11b and g solutions using directional antennas may still be viable, if the antennas are moved and the patterns changed in order to provide better coverage in the offices and minimizing out of building coverage. An additional downside of the 802.11b/g solution is the presence of many other users and services in the band that add to the noise floor and interference, and make reuse planning more difficult.

On the 802.11a front, the costs may be higher due to the limited coverage potential of any single antenna location, and the questionable ability to use remote antennas due to the excessive cable loss at these frequencies. On the plus side, less signal leaks out of the building and reuse planning is easier.

Since the IT department wanted a solution that minimized out of building coverage, needed little ongoing interference management, and would work well without worries about degradation due to uncontrolled interference from other users, it was decided to move away from a 2.4 GHz solution. This eliminated 802.11b/g from the decision.

It was decided to use a four AP per floor 802.11a solution. This eliminated routing unwieldy antenna cables, and assured that there was a significant amount of excess capacity for the future. The APs used omni antennas and were centrally located on each side of a floor, since this configuration gave the best coverage of the office and conference room space.

To accommodate the users' need for corporate network as well as public hotspot access, tri-mode 802.11a/b/g cards were selected as the client devices for the laptop users, thus allowing access to the company network via 802.11a as well as access to public 802.11b networks.

Although the coverage outside the building was de minimus, IT still worried about network security. Security issues were addressed in two ways. First, multiple key encryption was enabled on the wireless network. Since the wireless encryption standards are known to be breakable, this was only the first step in securing the network. The second step was isolating the wireless network from the corporate network. In essence, the wireless network exists as a standalone network that is separate from the corporate network. There is a VPN gateway at the edge of the wireless network that is used as the interface to the corporate network. A user accessing the wireless network is connected to the VPN gateway, and authorized as a legitimate user by this device. Once authorized, a VPN is created between the user and the corporate network. This VPN gateway is also used to secure all remote access to the corporate network via the Internet, including users accessing the corporate network from public wireless hotspots.

NEC, Fire, and Safety Code Concerns

Installing wireless LAN equipment in a building carries a few additional obligations that are often overlooked. If any of the cabling or equipment is to be mounted above ceilings or in walls, some jurisdictions require that the cable and hardware be either enclosed in metal conduit or be "plenum rated." Of course there is no way to enclose the hardware in conduit, so it must be fire retardant and comply with the appropriate standards, one of which is UL standard #2043. The manufacturer should note whether the equipment is compliant with this standard in the device's literature. Compliance simply means that fire retardant materials are used in construction of the cable or device. This is a very important consideration when mounting equipment above dropped ceilings that are used as part of the building's air return system or are otherwise viewed as common airways in the building. The reason for use of plenum rated cable and hardware or the use of conduit to enclose non conforming cable is that the use of the wrong cable or equipment in these areas could cause a localized fire to propagate throughout a large area of the building. It can also lead to the generation of large amounts of toxic smoke during a fire. That smoke could rapidly propagate through the building by way of the common air path available in the plenum.

Further there are very stringent National Electrical Code (NEC) rules for the routing of communications cabling. Areas that you should pay careful attention to are the rules concerning cable mounting and abandoned cables. Because of the in-depth requirements of the code, you should be familiar with all of Chapter 8. For example, although it is done on a routine basis, using zip ties to mount a Cat 5 cable is unacceptable and unlawful. There are specific guidelines for mounting and acceptable devices for mounting communication cables outlined in Chapter 8 of the NEC.

With most office buildings, each new tenant installs their own communication cabling and leaves the old cabling in place. This creates what the National Fire Protection Agency calls a "high fuel load," because the excess cable creates an increased fire and smoke hazard. In 2002 the NEC created new rules that require the latest installer to remove all abandoned cables from a building upon performing a new installation. For communications cabling the removal applies only to cable not enclosed in a metal conduit. This means that if there is unused Cat 5 in the plenum on a cable hanger it must be removed.

The NEC is still catching up with the increasing use of telecom cabling and equipment. Be sure to keep an updated copy of the NEC so you can be aware of future rule changes that will impact your installation materials and procedures. Ultimately, you as the installer are responsible for adhering to the rules and regulations, and it is the installer who bear the liability for damages caused by improper installation materials and practices.

Example Three: Community WISP

The factors defining a community WISP are about as numerous as the equipment choices available to serve them. With that in mind, I'll diverge from the previous examples and instead of discussing the specifics of a single system, I'll discuss several situations that might generally fall under the heading Community WISP, and look at some of the antenna placement, technology, and design alternatives available. As we shall see, each situation is quite unique and different from the other, and the applicable technical solutions vary according to the area and needs of the users. Additionally, these systems are most likely deployed as a business, meaning that there is an expectation of return on investment. The cost of solutions to serve these markets is as variable as are the technological solutions available for deployment. In analyzing the available technologies and selecting a solution you will often find several that seem to work well from a technical standpoint. In addition to finding a good technical solution, you will also need to consider the capital and recurring costs associated with each solution. Be sure to compare the expense associated with the base station equipment and the cost of the subscriber hardware. You will be deploying far more subscriber equipment than base stations, so the cost per subscriber served will be more greatly affected by subscriber equipment costs than by base station costs.

In addition, look at the ongoing operating expenses. If a technology needs more sites, then the lease costs and maintenance costs associated with the solution will be higher. It is useful to compare system alternatives by comparing the overall capital and annual operating expenses associated with each. By looking at the total cost of ownership on an annual basis, you get the best idea of which solution is most cost effective for your particular business. The accompanying CD-ROM contains a spreadsheet that analyzes the CAPEX and OPEX costs associated with a ubiquitous area wide wireless solution. While the complexity of such a system is much higher than the systems we are now discussing, the basic system needs such as equipment costs, lease costs, maintenance costs, interconnect costs, sales and marketing costs, and so forth are outlined within this spreadsheet. Reviewing the cost elements will assist you in identifying and capturing those costs associated with the business you are developing.

Community: A Garden Style Apartment Complex

For the first system let's take a look at a system that will serve end users in a garden style apartment complex. The users in this situation expect to have connectivity to their PCs in their apartments as well as the ability to take their laptops to the common areas of the community like the pool and clubhouse. In addition, some segment of the users would like the ability to use their wireless enabled laptop computers in other public hotspots.

From an RF standpoint, designing a system like this depends on the same basics as designing any other system: Knowing what power levels are available from the base station and client equipment and what antennas will be used allows you to determine the maximum tolerable path loss. Site surveys will show the attenuative properties of the construction materials used on site. Together, these factors will show how large the area served by each base station will be. Coverage area will then directly relate to the number of base stations necessary to cover the property.

The number of potential users per base station and the expected usage per customer will determine how much capacity each base station will need to serve. If the base stations have capacity demands that exceed the capacity available, then coverage can be modified by changing locations, antennas, or power output in order to reduce the coverage. Of course, making any of these changes will require a redo of the design, spacing base stations as appropriate to serve the demand instead of the coverage limits previously identified.

The customer requirement of mobility in and out of the community immediately limits the network design to using the prevailing standard for "hotspot" coverage. At the time this book is written, that standard is 802.11b. By the time you read this, it is quite possible that another standard will have displaced 802.11b. Even so, the system deployment guidelines will remain the same. You'll just have to make modifications to the coverage and capacity expectations in order to deploy the currently prevailing architecture.

RF Considerations

One of the first things that will need to be considered is the performance of the hardware to be used in the network. This includes not only the base station hardware but the client hardware as well. Because client equipment comes from a variety of manufacturers and has varying power output and receive sensitivity figures, the performance of the client card can become a significant factor in determining coverage of each base station. If, for example, you designed the system with the expectation that the client card had 100 mW of power output, then the coverage of the system would be set based on that expectation. If a customer already had a client card they wanted to use, but that card only produced 25 mW of output power, then the system may not work properly with this user. The 6 dB difference in power output would limit the service area associated with this user to distances half as great as those users with 100 mW client devices.

It is critical to determine a minimal acceptable performance level for client devices. If a potential customer has a device that does not meet or exceed this minimum performance threshold, then it should be replaced with a client device that does. This will assure that the system works well for all users, and that coverage issues are not the result of poor quality client devices.

Now that equipment performance has been considered, let's think about the challenges associated with covering this type of environment. In the last section of this chapter we discussed covering an office building. If we were now talking about a high rise apartment, similar design criteria could be used; utilizing equipment installed in common areas like hallways. Unfortunately, in the case of the garden style complex there is little common hallway area in each building and you cannot expect to gain access to individual apartments for deployment of infrastructure. So, what other common areas are available?

Are there attics or unutilized areas under the roof? If so, these spaces can become areas where equipment can be installed. Depending on how many floors exist in the building and how the buildings are constructed, placing directional antennas in attic areas with their main beam pointed down toward the occupied areas is one way of providing coverage to the tenants. Another is to mount omnidirectional antennas on the roof. Note that it is important to make sure that you use low gain omni antennas or that there is considerable electrical downtilt implemented

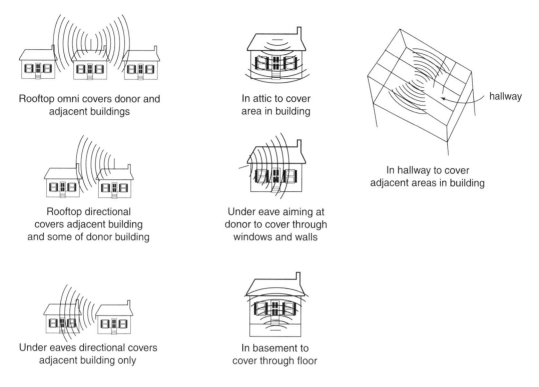

Rooftop omni covers donor and
adjacent buildings

In attic to cover
area in building

hallway

In hallway to cover
adjacent areas in building

Rooftop directional
covers adjacent building
and some of donor building

Under eave aiming at
donor to cover through
windows and walls

Under eaves directional covers
adjacent building only

In basement to
cover through floor

Figure 6-1: Antenna installation options

in high gain omni antennas. If not, the main beam will shine over the intended coverage area, leaving poor system performance in the apartments and (probably) excellent coverage in adjacent areas.

Another option for coverage is to point directional antennas so they are penetrating the least attenuative parts of the building: the windows in the outside walls. This deployment would require directional antennas placed on or under the eaves of the roof, and either pointed at the windows and walls of adjacent buildings or pointed down and back at the windows and walls of the building to which they are attached.

The selection of the best location and antenna mounting options will be driven by the real-world variables associated with the particular buildings to be covered. If the buildings have a stucco exterior, you may find that the wire mesh to which the stucco is affixed causes too much attenuation. In this case trying to "shine through" the sides of the building may not be an appropriate solution. On the other hand, there may be a sufficient area comprised of windows, in which case the propagation through the windows will be sufficient to provide service to the user.

If the buildings are single or two story and are not built like concrete bunkers (in other words they have traditional drywall and wood floor construction instead of poured concrete floors and walls) then locating the base station equipment in the attic space shining through the ceiling may be the better solution.

Another consideration that can have an effect on your antenna placement decision is capacity. It is possible that the location giving the best coverage actually provides so much coverage that a single base station does not have sufficient capacity to serve the customers in the covered area. In this case, selecting a "poorer" antenna location in a location providing less coverage may provide a solution that is more in line with balancing the coverage/capacity needs of the system.

In the end, the only way to know will be to do a site survey using antennas mounted at various spots that seem reasonable from a coverage standpoint. By learning more about the attenuative effects of the construction materials used, and which areas of the building offer the least attenuation to the signal allows you to rank the designs according to which best covers the desired area. Of course, the best coverage design may have intractable problems from an implementation standpoint, so knowing what solution is second or third best is useful in case compromises are required in order to install the equipment cost effectively.

The best coverage locations may also have a problem with interference management. Because the covered area is large, many base stations will be needed. Since the system we are discussing is 802.11b-based, it will definitely require frequency reuse in order to have sufficient AP locations to provide full coverage of the community. Again, the site survey will help identify which locations are the best compromise for gaining sufficient coverage without allowing interference to become unmanageable.

Interference management is made more difficult in a situation like this one if the antennas are located outdoors with the intent to cover indoor areas. A strong signal is required to overcome the attenuation of walls and windows, internal walls, and obstacles. The RF design will have to allow at least an extra 10 to 20 dB of link margin in order to overcome these losses and provide useable signal inside the buildings. Unfortunately, this extra attenuation does not exist everywhere, so the signal outside can carry significantly further than the intended coverage area. Because it propagates further it can be seen by other APs on the property and, potentially, users far from the area where it provides primary coverage. The extent of this effect is shown graphically in Figure 6-2.

Backhaul

Because of the size of the covered area and the number of buildings and APs, a method of networking the system together will be needed. In addition, connectivity to the Internet will be required. Depending on the location, the Internet connection can come from a number of sources. Local RBOC, CLEC, or ISP operators may be able to provide a wired connection with sufficient speed to meet the usage demands of the system. Alternately, there may be a wireless ISP in the area who can provide connectivity.

Once this connectivity is secured, it will be brought to a single location on the property. The system implementer will need to extend this connectivity around the area so the individual APs can be connected together as a network.

Multiple APs in each building could use CAT5 or similar cable to bring the individual data drops to a convenient central location in the building. An Ethernet switch or hub can be used as a traffic aggregator at this location. Using CAT5 cable between buildings is probably not feasible for several reasons. First, CAT5 runs are limited to 300 feet, which is probably insufficient length to connect two or more buildings. The second hurdle is finding a way to route the cable between the buildings. Unless unused underground ducts exist, there is no practical way of routing cables.

A better alternative is the use of wireless for backhaul. Do not use radio equipment operating in the same frequency range as the equipment providing end user connectivity. Doing so only increases the amount of channel reuse and attendant interference levels. Since this example is using 802.11b gear operating at 2.4 GHz, the backhaul hardware could operate under the 802.11.a standard in the 5.6 GHz band. Use the building where the Internet facility is terminated as a donor station by installing an 802.11a AP and an omni antenna of appropriate gain in a location where it can be seen by each of the surrounding buildings. Each

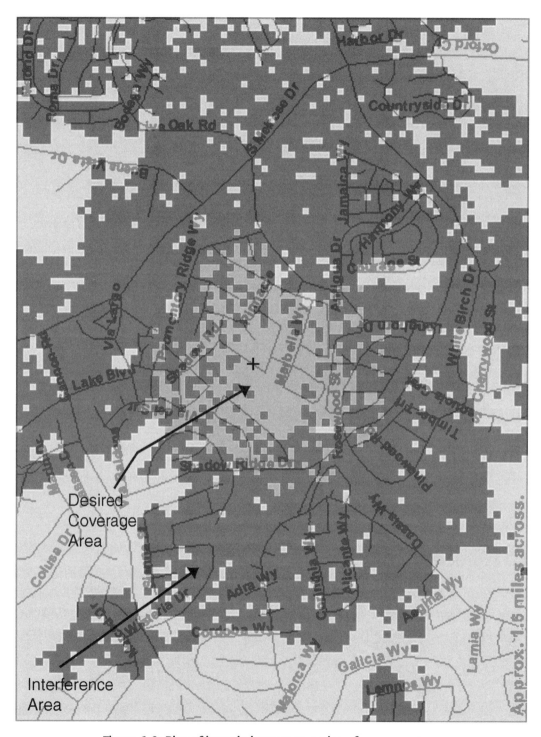

Figure 6-2: Plot of intended coverage vs. interference contour

surrounding building would have an 802.11a client device and antenna mounted so that there is a clear line of sight back to the donor station.

If a single AP cannot accommodate the traffic backhaul requirements on the property, then use multiple 802.11a APs with directional antennas in order to serve individual buildings with higher bandwidth. A separate channel should be used at each AP in order to eliminate interference and channel contention issues.

Weatherproofing

Because of the low power nature of 802.11 devices and the loss associated with coax at frequencies over 2 GHz, mounting the APs at a central location inside and using coax to connect to the antennas will not be a feasible idea.

This means that the outdoor antenna locations will require radio hardware that is weatherproof. This can be accomplished by purchasing equipment intended for outdoor use, and therefore manufactured with weatherproofing in mind, or using indoor equipment and installing it in a weatherproof enclosure. Allowing the equipment to be mounted as close to the antenna as practical reduces the coax loss to a minimum, and assures maximum power output and receive sensitivity. This brings up another area that requires attention: the cable used in the deployment. Make sure that all coax cable, power cable and CAT5 cable is rated for outdoor use or that it is enclosed in conduit. Further, local building and safety or fire codes may require the use of Teflon or other fire resistant materials as the jacketing material of the cables.

Also, make sure that all antenna and feedline connectors are properly sealed and waterproof. Even in dry climates improperly sealed cables can be damaged by water because they "breathe" due to changing temperature. The cable heats during the day, expelling air. In the evening as the cable cools, cooler moisture-laden air returns to the cable. Over time this leads to condensation collecting in the cable and connector. This condensation changes the characteristic impedance of the cable, and can lead to reduced power output and apparent receive sensitivity due to the growing mismatch of the coax impedance and the antenna.

Grounding and Lightning Protection

Proper equipment grounding and lightning protection is also a consideration when mounting hardware in outdoors locations. Because the antenna is mounted at a high elevation, it becomes a prime lightning strike candidate. A lightning strike can damage equipment connected to the antenna or, in the worst case cause a fire inside the building where connected hardware is located.

Lightning protection relies on a good ground, so let's discuss grounding first. Grounding, like the name implies, involves connecting the equipment to an earth ground in order to discharge any static electricity collecting on antennas and lines, or to provide a direct low resistance path to ground for a lightning strike. This last point: "a low resistance path" is a critical element in defining a ground. The ground itself must provide as little resistance to the earth as possible. Driving a 4 foot long copper clad rod into the earth does not make a good ground! A good ground is composed of multiple 8 foot minimum length ground rods spaced at intervals around the base of the tower or building with each rod welded to a common heavy copper ground wire. Often, these grounds are completely buried and inspection ports are provided at the points where the rods are located.

The common copper wire from the ground connects to the tower legs and to a copper buss bar at the base of the tower called a *ground window*. The ground also extends to another ground window located inside the equipment room. The ground window on the tower is used as a point where the shields of all coax cables and an independent ground wire from each antenna on the tower can be grounded. It also provides one spot where the inline lightning protection installed on the coax can be grounded. The ground window in the equipment room provides a grounding spot for equipment racks, equipment, lightning protection, and power supplies.

It is important that all grounds at the site connect together to a common ground. This means that telco and electric utility grounds must also be bonded to the ground window. If not, the differences in resistance of independent grounds can cause voltages to flow between equipment connected to the different grounds. This will cause damage during lightning strikes, because part of the lightning energy will flow through other equipment in search of the path of least resistance to ground.

Lightning protection devices are designed to protect a number of devices from lightning strikes. They can be purchased to connect in line with coax, to protect the radio equipment connected to the antenna, they can be sourced in RJ-11 or RJ-45 configurations and used to protect telco and data equipment. No matter what the form factor, these devices all provide a common service: they provide an alternate low resistance path to ground during a lightning strike, and thereby protect the electronic equipment connected to them. That's why it's important to have an excellent site ground. If the site ground is inferior, the lightning protection cannot offer the low resistance ground path, so the lightning will discharge through the electronics, leading to damage.

If you are locating antennas and equipment in an existing site with other communication carriers, the site ground is probably good. If you are the first on a site, or are developing your own site, get a professional to evaluate any existing ground and make recommendations for improvements to existing grounds, or to design new ground facilities to accommodate the local soil conditions.

Community: A Small Area Subdivision

This community is comprised of high density residential housing. The area to be covered measures one square mile and contains 300 homes. There are currently no other broadband facilities available in the area, so the wireless system will provide the first and currently only broadband Internet connectivity available.

Equipment Selection

The equipment selection and implementation decisions become significantly more complex when trying to serve an area this size. From the equipment standpoint, while traditional low power 802.11 products can be made to work in this environment either by deploying many of them around the neighborhood or using fewer units with high gain antennas located at high elevation locations, they will never perform as well as a product designed to function in this environment. High power solutions using smart antennas, like the Vivato product, can offer improvements in coverage. Depending on the particulars of the area, it might be covered with two to four Vivato-type products. Still, Vivato is 802.11-based and carries the limitations of 802.11 technology.

As previously discussed, 802.11 uses CSMA/CA for sharing access to the system. Because 802.11 was designed for LAN replacement, it expects all users to be able to hear each other. In the case of a community spread out like this you cannot expect this to be the case. In an effectively designed system, there will be many users who have great connections to the base station but cannot see each other. Additionally because there is no power control in 802.11, there is a "near-far" issue in 802.11. This problem manifests itself where there are users very close to the base station, and other users far away. The close-in users present a significantly stronger signal to the base station. This stronger signal can completely override the weak signal coming from a distant user.

Invoking RTS/CTS and setting low packet fragmentation thresholds can improve the situation to some degree, but this has the side effect of significantly decreasing system capacity.

It is also not realistic to expect the service to be directly useable by an 802.11 client card in a computer located inside the home. While there will be certain homes where this is possible (those close to the base station with an unblocked path), the majority of properties will not be covered with sufficient signal strength to guarantee such service. This too could be remedied with more equipment at more locations, however the cost of deploying, operating, and maintaining such a network will be prohibitive.

There is one final consideration to keep in mind. The coverage of a system will only be as good as the lowest performance radio hardware in the link. Having a high power AP will not compensate for a low performance CPE device. When using high power AP equipment in community coverage situations, it is critical to assure that all subscribers use CPE that has power and receive sensitivity commensurate with that of the AP. Otherwise, performance and coverage will suffer.

A better solution, if the community Codes, Covenants, and Restrictions (CC&Rs) allow it, would be to use externally mounted CPE radios and directional antennas at the customer's home. These small devices would be mounted in such a way as to clear the roof and be pointed toward the base station. This extends the service range of each base station by adding gain at the client end and improving the path loss characteristics of the link.

Locating the Customer Premises Equipment (CPE) outside or in a window facing the base station will improve the coverage potential of the system whether it is 802.11-based or based upon another standard or proprietary solution. Decreasing the path loss always improves the stability, throughput and coverage of the network.

Since 802.11 is not well suited to this type of network, other technologies should be considered for use. There are several proprietary solutions available that are designed for this environment. Also, the 802.16 standard was designed for use in this type of environment. The MAC and PHY layers in 802.16 do not suffer from the limitations of CSMA/CA and the near far problems associated with 802.11.

From an RF standpoint, deploying 802.16 or a proprietary solution in this location is no different from deploying 802.11. You need to know the frequency of operation, the power output, receive sensitivity, and antenna gains of both the base station equipment and the client equipment. With these parameters you can use the guidelines and tools presented in this book to determine the allowable path loss and estimate coverage from each base station.

System Planning

Once you've narrowed the field of hardware solutions, it's time to take a close look at the environment to see where the best equipment locations will be. In particular, look for high elevation areas (hills) or objects (taller than average structures) in or immediately surrounding the area to be covered. The more you can raise the antenna above the local clutter, the better your coverage will be.

Once these areas have been identified, run a propagation model to determine which are the best choices. With a ranked list of choices in hand, it's time to conduct site surveys. The purpose of the survey is to identify ownership of the property, availability of the property for your use, identify any zoning restrictions that would disqualify the location from use, and identify if any of the locations have availability of high speed Internet connectivity either through an RBOC, CLEC, or ISP. If none have such connectivity, you'll have to continue looking for a site that does have connectivity as well as line of sight paths to your target coverage sites. This additional site will be used as the hub site for aggregation and backhaul from the traffic carrying sites.

Having identified locations that are available for your use, an RF survey should be conducted. As with all surveys, the purpose is to assure that the selected hardware does indeed provide the coverage that you expect, and that the location can support the selected traffic backhaul to the aggregation point in the network. Once you are satisfied with the performance of all locations needed to cover the area, you'll need to arrange for leasing and zoning at the selected properties. Once leased and zoned, installation of the network hardware can begin.

Community: An Urban or Suburban Area Serving Business Users

In this example we move away from the residential consumer market. In this case we are looking at a high density business setting such as would be found in a large office park or in an urban or suburban commercial zone.

This system will provide very high speed facilities (>10 Mbps) to buildings. The distribution of bandwidth to users in the building will be accomplished with a network using traditional CAT5 wiring. Because of the large bandwidth requirements, the system will need a hub location that has access to high speed wired facilities. The most obvious choice would be a building with an existing connection to a fiber optic network.

The wireless network in this case is used to extend the connectivity of the fiber backbone to buildings that are too far from the fiber to be cost effectively connected by extending a fiber connection to the building in question. In many urban locations, the cost of extending these connections can be over $1 Million per mile. This leaves wireless as a much more cost effective solution in these areas.

Spectrum Issues

Because this is a system serving commercial interests, security, reliability, and uptime will be a much more critical element than they were when deploying a consumer Internet access network. These factors alone should be enough to drive you away from using Part 15 spectrum as the basis for the network.

So now we are looking at licensed spectrum and equipment designed to work in these allocations. 802.16 deployed as a point-to-multipoint network may work in this environment. Individual point-to-point microwave links may also work. In fact, depending upon the bandwidth requirements of each building, a system combining both alternatives may be appropriate.

Availability of licensed spectrum below 5 GHz is rare. It is still possible to coordinate and obtain licenses for 18 GHz and 23 GHz point-to-point microwave links from the FCC. In addition there are several companies who own 38 GHz spectrum and are making it available for lease to third parties.

Links at these frequencies do have limitations. They must be Line of Sight (LOS). In a point-to-point deployment they require a separate antenna for each link, which may cause a proliferation of dish antennas on the hub site. Rain fade, the additional path loss associated with rain falling across the radio path, also becomes a problem as the frequency increases. Links must be kept short (under several miles depending on frequency and area) in order to maintain reliability. In fact, because these links are LOS, and therefore the path loss over distance is a known factor, most manufacturers of equipment for these types of systems can offer guidance on the average link lengths supportable by the area you are planning to build. The reason link lengths are area specific is because of the effects of rain fade. Above 10 GHz, rain fade becomes a design consideration because heavy rain causes additional attenuation of the signal. As the operating frequency increases, the deleterious effect of rain fade also increases. The amount of rain fade is dependent on the intensity (inches per hour) and duration of these peak rainstorms. Luckily, the physics of weather and the intensity and duration of storm cells have been well characterized around the world, so tables exist which identify the rainfall characteristics of an area, the extra loss associated with this rainfall, and the additional signal headroom you will need to design into the system in order to maintain a particular link uptime.

On the plus side, there is plenty of channel bandwidth at these frequencies, so extremely high bandwidth (over 650 Mbps) can be achieved by networks utilizing these frequencies.

The 18 and 23 GHz bands are limited to point-to-point links, and because of these rule limitations will be unusable for an 802.16 system deployment. The 38 GHz spectrum rules do support point to multipoint system deployments, so it would be possible to deploy 802.16 in the 38 GHz band.

Design Considerations

As you can see, the availability of spectrum and the rules governing that spectrum will drive the selection of technology. The bandwidth requirements of the end users will dictate the method used to deploy the network. For example, if all buildings have extreme bandwidth requirements, then a point-to-point network built as a hub and spoke network may be most appropriate because it allows the maximum bandwidth per link. In such a network it may be important to provide physical redundancy of the radio hardware in order to assure that a hardware failure does not cause service outages. This is easily achieved; however it is very costly since it doubles the radio equipment requirements.

If the bandwidth requirements are small enough that several buildings can share the capacity of a single radio, then you could consider point-to-point facilities constructed in a ring architecture. This reduces the cost of outage protection by using the ability to transmit traffic in either direction around the ring. A single break will not isolate any user, but may impair the traffic handling capability of the network until it is repaired. Because of this "route redundancy," the ring does not need the equipment redundancy provided by backup radios on each hop. Instead it requires only one additional radio: the one that connects the last site in the chain back to the hub site, thus completing the ring.

If appropriate spectrum is available, an 802.16 or similar point-to-multipoint network could be considered. This network would be designed similarly to the hub and spoke network. For all practical purposes, the architecture of the two solutions is identical: Both depend on a central facility that originates and terminates traffic from a number of far end stations. The architectural difference between the two boils down to the antenna system on this central site. Instead of using dedicated antennas for each link, point-to-multipoint networks, including 802.16 systems; use a single antenna to connect multiple far end stations. These antennas could be omnidirectional or directional depending on the needs of the design. The decision on antenna pattern will be driven by the gain required to make the link, the area to be covered, and the capacity requirements of the area to be covered.

In either case, the RF design is simplified because you only need to consider a fixed number of end stations, each of which has an easily calculable path loss due to the fact that they are all line of sight paths with known rain fade requirements.

Your design will need to focus on assuring that the buildings selected do have line of sight paths, or that you can acquire intermediate locations which have line of sight and can be used as repeater locations.

Community: A Small Town System for Consumer and Business Users

This system design is a logical expansion of the past two community examples. There are many small towns across America that have limited broadband connectivity. A wireless network can effectively serve these communities. The community is made up of both residential users and business users, each of whom have specific needs we have discussed in the previous sections of this chapter.

This is another situation where an 802.16 network would be an ideal solution. With its ability to offer Quality of Service levels and varying speeds, it can provide a good match to both residential users who are looking for a DSL like connection and to the business user who requires a stable, high throughput connection that can offer consistent throughput performance. Such a system could operate in licensed or unlicensed spectrum, however the licensed band hardware will have the ability to serve larger areas because of the higher allowable power output.

There are also some proprietary solutions that could work well in this environment. The Motorola Canopy™ solution and the Proxim Tsunami™ multipoint are two solutions that use unlicensed spectrum. The Navini RipWave™ hardware can be purchased in both unlicensed and licensed versions.

Once again, the key to cost effective coverage will be finding the locations in and around the area that offer the best line of sight paths to as many locations as possible. Tall downtown buildings are an option as are towers located within the coverage area or on nearby hills. Propagation models and site surveys will assist you in selecting the best sites. Backhaul could be provided by wireless or wired facilities depending on what is available and cost effective.

Summary

The examples we've reviewed show that there are always multiple choices. Trade-offs will need to be made based upon the particular needs of your customer and business. Certain technical solutions may offer low cost of entry but that benefit may be eroded by higher ongoing operating costs associated with the system. Others may have inexpensive base station hardware but high cost subscriber terminal equipment. Still others may be inexpensive but limited in their ability to meet the needs of the customer. Ultimately, these are the trade-offs you will need to make by balancing the capabilities of the equipment to the financial needs of the business and the needs of the customer. In many cases there will be several solutions that could work well, each with its own benefits and shortcomings. When in this situation, it may be useful to look at others in the same line of business and see what they have selected. If a particular solution has been adopted by a number of disparate companies, there may be good reason.

Example Four: Mobile Broadband Network

So we finally take a look at the Holy Grail of wireless data systems: a wide area ubiquitous broadband network capable of providing >2 Mbps of bandwidth to mobile users traveling at highway velocities. In addition to providing mobile coverage, such a system will also by default have the ability to provide service to fixed locations, and could become a real competitor in the broadband delivery business by competing against DSL and Cable Modem solutions.

Because of the need to assure ubiquitous coverage and depending on the area to be covered and the frequency at which the system operates, such a network will require thousands if not tens of thousands of base stations, and cost hundreds of millions of dollars. Higher operating frequencies will require more base station locations due to propagation loss increasing with frequency.

In order to be effectively deployed, such a system will require licensed spectrum in a band under 5 GHz. It will probably also adhere to a standard like CDMA2000, 802.16e or 802.20. Unfortunately, at the time this book is being written, none of these standards (other than the low speed first generation CDMA2000 solutions of 1XRTT and 1XEVDO) have been finalized and there is no available equipment conforming to any of these specifications.

Initial Modeling

That need not stop us from considering what it will take to design and deploy a network. In fact, that's what the MSA and RSA Operators models on the

CD-ROM are used for. These models take a high level approach to determining the quantity of equipment necessary to build the network, the number of base station locations to provide coverage and capacity, the expected subscriber growth over time, the capital expenses associated with the business, and the operating expenses associated with the business. While they are not accurate enough to use to actually design a network, they are useful for financial planning and comparisons of technology and frequency variables. Using these models, or another like them, is the first step in understanding the magnitude of the system and its costs.

This is only a first step. The granularity of the information gleaned from the use of such a simple model will not be sufficient to actually build a network. After we review the preliminary business planning tools, we'll discuss the tools and steps necessary to expand the conceptual design into one that could actually be constructed and operated.

Preliminary Information

In order to begin planning this network you will need to identify the area to be covered along with the percentage of the area considered urban, suburban, and rural, the frequency of operation, and the technology to be deployed. The technology being deployed will give you the power output and receive sensitivity associated with the hardware so the basic link budget can be determined. In addition, you will know the capacity of the hardware. This will be useful when you begin analyzing the capacity requirements of the network based on user demand and the demographics of the area. This allows you to determine if a single base station location can meet the traffic demands in the area defined by its maximum coverage.

In addition to the technical specifications, you will need the rough demographics and morphology (land use characteristics) of the area, including the population and its distribution throughout the urban, suburban and rural areas, plus the business plan expectations of subscriber growth and average usage characteristics. You will also need to estimate the costs of all hardware, services, and ongoing expenses (such as lease costs, facility costs, and so forth).

Coverage Modeling

With this basic information, you can begin to model the system. First, determine the link budget of the equipment. Now that the Maximum Allowable Path Loss (MAPL) is known, it can be used as an input to whatever propagation model you wish to use for analyzing coverage. The operators model spreadsheet uses the well known COST 231 HATA model. This model allows for the characterization of NLOS loss in a variety of environments and so can model coverage behavior that is different for sites in urban areas vs. rural areas. The other input to the propagation model will of course be operating frequency.

The output of this modeling exercise will be the average coverage radius you can expect from any site in each morphological area you've defined. From this radius, the diameter and subsequent area of each site can be determined.

Since the area to be covered in total, as well as its morphology is known, you can now determine how many base stations are needed in the area using simple math (area to be covered / per BS coverage area). You now have a good estimate of how many base stations will be necessary to serve the area. The next step is to determine if the aggregate capacity of those sites is sufficient to meet the usage demand in the area.

Capacity Modeling

Here is where demographics become valuable. You need to know the population of an area and its distribution over the different morphologies in the market so you can determine how much of the population is contained within the coverage area of each base station.

The total population per base station is then multiplied by the percentage market penetration expected. This identifies the number of subscribers per base station. By multiplying the average usage per subscriber by the number of subscribers per base station you now know the demand that will be placed on each base station. If the demand exceeds the capacity, additional base stations will need to be added to the system in order to increase the total capacity of the system enough to meet the expected demand.

Congratulations! You've now estimated the total base station quantity necessary to serve the market. With this number you can calculate the cost of the network.

Cost Modeling

Largely, the number of base stations deployed drives the network cost. In addition to the cost associated with each base station, there are costs associated with support hardware at each site like power, backhaul, antennas and cables. There will also be costs associated with aggregation facilities where traffic from a number of sites in an area is collected and concentrated before being sent to the main routing center in the network.

In addition to capital expenses, there are ongoing operational expenses associated with each base station. Lease costs are a recurring expense at sites that are not owned and constructed by the operator. The total number of base station sites also drives maintenance costs and facility costs. The financial (CAPEX and OPEX) inputs worksheet in the operators model spreadsheets breaks down these costs, and provides you the opportunity to use your values to define these variables.

Based upon these inputs, the model calculates the system capital and operating costs, which can be used as inputs to the rest of your business plan, just as they are used to provide summary output in the operator's models.

Designing in the Real World

The previous design walk through was useful as a business planning exercise. It is only marginally useful in designing a real system. The difference between the planning exercise and the real network design is a matter of granularity. In the planning exercise, there was no reason to know exactly where a site was located, or exactly how the population and traffic was distributed over each site.

In order to design a system that will actually be built, these questions and others must be accurately answered. To start with, a real design will be based upon the actual coverage achievable from each base station and an accurately estimated number of users within the coverage area. Neither the simple spreadsheet tool, nor the RadioMobile software discussed earlier, is sufficient for this task. A computer modeling tool that includes terrain, morphology, accurate predictive algorithms, and fine resolution demographics will be needed for this exercise. Such tools are available from a number of sources like LCC Inc., MSI Inc., EDX Inc., and WFI Inc., as is the engineering talent to use them and assist in designing and building a network. Just as it was necessary in our simpler networks, it will be necessary to evaluate the predictive models ability to accurately determine the propagation behavior of the area to be covered. The experience of the purveyor of these models will be extremely useful in tailoring their model to your needs.

The demographics that are necessary must be based upon the smallest area available. That could be government census data organized by postal code and further distributed by square kilometer or mile. The demographics should be broken into categories that will be helpful in defining a customer. Such factors as age, income, education, road miles, and businesses are used by marketing organizations to define the makeup of a subscriber. By applying these same weightings against the demographic data available for the model, individual sites will more accurately reflect the true number of users rather than a market average.

As you begin and plan an actual system, the first challenge will be to identify available properties that can be used as base station locations. The propagation of these locations is modeled, and matched to the underlying demographics. The total number of possible users in the sites coverage footprint will be output along with the modeled coverage of the site. By evaluating the expected market penetration percentage and the usage per subscriber, the model will predict the expected number of users covered by the base station and the total demand placed upon it.

These complex commercial models allow you to add all the sites that make up a network, and then to see graphically whether the coverage and capacity are sufficient. If there are problems, sites can be added or moved until the network is optimized for coverage and capacity. In addition to coverage and capacity, there are frequency reuse considerations. These models also consider and calculate the effect of co-channel interference, and can help to optimize a reuse plan acceptable to the C/I requirements of the equipment you are using.

These models can also be used for planning radio-based backhaul networks. By analyzing the line of sight opportunities between different sites, they can assist in the design of a point-to-point or point-to-multipoint backhaul network.

The output of this modeling and planning effort will be actual address, longitude and latitude of sites, the height and type of antenna, the number of radios necessary to serve the demand in the area, and the power output of each base station. These outputs will be used for acquiring real estate and constructing individual sites.

These tools will continue to be used throughout the life of the network. As the number of subscribers and usage increase, the network will need to be expanded to meet these growth demands. The modeling tool can be used to find optimal locations for new sites and to frequency coordinate channels for these new sites. They can also be used for forecasting system requirements based on anticipated growth, so they remain a valuable tool for planning the annual growth requirements of the network based upon the subscriber growth and usage predictions of the sales and marketing staff.

Chapter Summary

As you see from the examples presented in this chapter, there is no "one size fits all" solution. There are many unique situations and business opportunities, as well as many different technical solutions. In some cases there will not be a clear advantage to one technology or another, in others a clear "winner" will emerge from your analysis. The more you understand the environment you are trying to cover, the expectations of the business, and the users of the technology, the better chance you will have of selecting and deploying technology that meets the needs of all parties.

CHAPTER 7

Back Office System Requirements

- Network Systems Required
- Customer Authorization System
- Billing Data Collection
- Network Monitoring and Control
- Billing System
- Trouble Ticketing
- Customer Service Systems
- Design Considerations and Requirements

Back Office System Requirements

Throughout this book I have either explicitly or implicitly discussed cost as a part of the system design. Even if the network is a private one, costs need to be considered in order to assure there is adequate value received for the expenditures made. In the case where you are planning to construct area wide WISP coverage or deploy a mobile coverage network, you most likely are doing it for profit, so beyond just managing CAPEX and OPEX costs, you need some way of managing subscribers, billing them and collecting their payments. In addition, if you are billing for service, the customer has some expectation of service reliability. To assure this, there is a need for a method of monitoring the network and remotely facilitating fault isolation.

The software solutions that perform these functions are varied in their capabilities and complexity, and are known collectively as *back office systems*. This chapter will provide an overview of the types of systems needed to support a business. While a full-on metropolitan mobile network will require all of the subsystems discussed, a local WISP with a few hundred customers will be able to make do with significantly less complexity, and may be able to handle the majority of the tasks without significant levels of automation.

Figure 7-1 illustrates the various types of back office systems and their interface and interaction with each other and the network. As you can see, the back office system is actually a multitude of subsystems each dedicated to a specific task, but needing the cooperation of and communication with most of the other back office subsystems and the hardware comprising the network.

Network Systems Required

The systems described in this chapter may either be part of the system solution provided by the equipment vendor, or these systems may be comprised of third

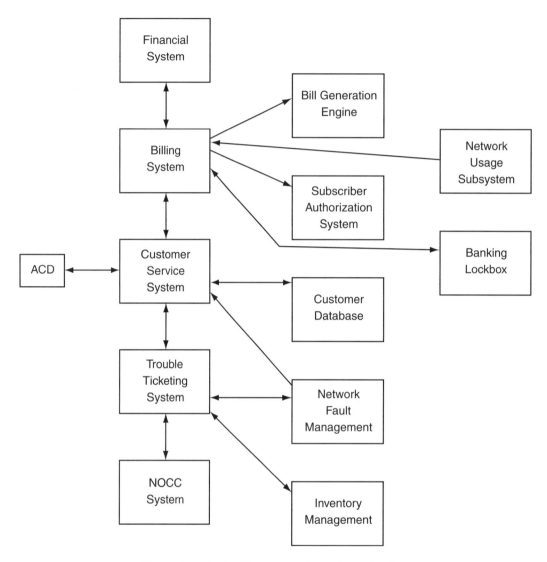

Figure 7-1 Back office system interdependencies

party hardware and software designed to interface with the equipment and provide functionality not otherwise available in the equipment as initially provided.

Customer Authorization System

Since this system is offering service for hire, it needs to deny access to anyone who is not a paying customer. A RADIUS server is one way of accomplishing this. Other systems may have other solutions, but the basic requirement for this system is for it to act as a gatekeeper, identifying and allowing access by legitimate users and denying access to others. This screening can be accomplished by requiring a user name and password when logging onto the network, or it may use a MAC address or other unique identifier associated with the subscriber's equipment. This system interfaces with the billing system so customer connects and disconnects can be handled.

Billing Data Collection

If the service being sold has a usage sensitive component, then a system that monitors and collects data about the number of packets or MB of information used by each customer will be necessary. These systems are normally located at the central interconnect points of the network, and "sniff" the traffic crossing the facilities. By monitoring the traffic, these systems can collect information about all packets sent from all IP addresses. Since each concurrent user has a unique IP address, this usage information can be mapped back to the IP address "owner" at the time it was collected. This usage information is collected in a database, which is used by the billing system to prepare a usage-based bill.

Network Monitoring and Control

Another important system is the Network Monitoring and Control (NMC) system. Sometimes called an NOCC, or Network Operations Control Center, this centralized system provides a single point for monitoring all the hardware that comprises the network. This can be accomplished using a combination of SNMP commands to equipment that supports SNMP, or by proprietary interfaces to the hardware, supplied by the manufacturer for the specific purpose of remote network management.

Often the functionality of the NOCC is comprised of numerous subsystems that all feed information to a high level graphical user interface (GUI). This GUI is what network technicians will interact with when monitoring or troubleshooting the network.

Another subsystem is the trouble ticketing system. This system must be able to take trouble reports from customers as well as trouble reports that are internally generated by the monitoring system. The trouble ticketing system is a two way system that should be able to forward the trouble report to the technician responsible for the area in which the trouble was identified as well as providing a way for the responsible technician to input information necessary to close the report when the fault is isolated and repaired. Both the trouble ticket and the corrective action are kept in a database. This data can be useful in identifying recurring problems or system weaknesses.

Billing System

From its name, you'd expect this system to be nothing more than the financial Accounts Receivable system. In reality it's much more complex. A good billing system handles all customer to system transactions from sign up, to interval billing, to customer service record keeping, to automated service disconnect for non payment.

A good billing system can become the core engine that drives customer interaction. The billing system will be one of the few systems that the customer actually comes in contact with. From the moment service is requested the billing system begins handling certain transactions. For example, how is the new service provisioned? Does it require a customer service or sales person to interface with the billing system to establish service, or is the system customer friendly and on line thus allowing the customer to self provision the service? Either or both of these options may be the right one for your business, and there are systems available that support both options.

Once the billing system has the basic information about the customer, things like name, address, contact information, billing and payment method, and identification information about their hardware, it can update several subsystems. Customer name, address and contact information become part of the customer record used by other organizations such as sales, marketing, and customer service. The billing and payment information becomes part of a billing record in the subsystem that manages customer billing and collection. This system may generate paper billing statements that are mailed to the customer, or it may generate e-mail billing statements and automatically send them to the customer. It may even be set up to do a recurring credit card transaction against the credit card number supplied by the customer.

In order to accomplish such things as customer setup and billing, the billing system needs to be interfaced to the network elements in such a way as to give it the ability to automatically provision or disconnect service to a customer, and to near real time monitor the usage of a customer so that usage sensitive billing can be offered. So, a connection to the RADIUS or other authorization system is required, as is a connection to the network usage system that keeps track of individual users connectivity and use of the network. This connection allows not only fixed monthly access to be a billable event, but also allows additional charges per MB of traffic sent and received. Depending on the complexity of the usage information being collected, it could also provide tiered billing allowing higher costs to be associated with real time QOS transactions like VOIP or streaming video, and low costs associated with non real time events such as ftp transfers.

Another critical capability of the billing system is its ability to manage multiple pricing plans, including short term promotions and specials. The billing system is a computing system, and therefore can only allow those price plans that are programmed into it as options for a customer. If the complexity surrounding adding a new plan is too high, then the billing system may cause the company to become uncompetitive in the market because of its inability to quickly respond to a dynamic marketplace.

The billing system must also have links to the banking system used. If payment by check or cash is accepted then there will most likely be a lockbox relationship with a banking institution to manage the collection of these checks. As the payment arrives at the lockbox, the payment information must make its way to the billing system as an update to the current account status. Similar links are needed for managing credit card payments, and additional relationships with a financial institution are needed if automatic charges or debits to a credit card, debit card, or checking account are contemplated.

Trouble Ticketing

In order to manage the maintenance of the network, a method of generating and closing trouble reports is needed. This system derives its information from several sources. Customer trouble reports collected by customer service, which appear to be network related, are sent to the trouble ticketing system. Also, known troubles identified by the NOCC are sent to the trouble ticketing system.

The trouble ticketing system can identify trouble reports by type or by area, and use this information to identify the individual technician responsible for investigating the issue, and forward the trouble report to that specific individual. As with all these systems, the degree of automation is driven by the size and complexity of the network.

The first step in trouble resolution is to identify that a problem exists. This is initiated by a customer reporting a problem with service or by the system reporting a hardware failure. This event opens a trouble ticket. The ticket is assigned and forwarded to the tech responsible for resolution, and a clock is started. The clock runs until the tech reports back that the problem is resolved. During this time the system can report ongoing status of the trouble ticket, such as open assigned, open pending forwarding, open and being worked, and so forth. When the problem is resolved, the tech closes the ticket by connecting to the trouble ticket system and reporting what was done, and if necessary, what hardware was replaced. A copy of the closed ticket is sent to the originator to inform them of the resolution. In addition, if hardware was replaced, the serial numbers of the defective hardware and the serial number of the replacement hardware are forwarded to the Inventory Management subsystem. This allows all equipment to be tracked to its current location thus assuring accurate inventory management and control.

The trouble ticketing system keeps a database of all reports, who cleared them, what was done, what equipment was used, and how long the ticket was open. This information is valuable for future review, and can identify common equipment problems, coverage or interference problems, design weaknesses, and personnel training issues.

Customer Service Systems

Customer service organizations can be subdivided into three categories: Those supporting activities associated with subscriber service initiation and change, those supporting subscriber trouble reports, and those supporting subscriber technical support. The systems supporting these unique functions are similar in their need to connect to other back office subsystems. The key difference is potentially only the training level of the personnel handling the request. Activating a customer requires far less skill than assisting a customer with a troubleshooting and problem isolation process.

Contact by the subscriber can be made real time using the telephone, or off-line using e-mail or web-based request forms. The customer service system therefore needs not only connections to the other back office subsystems, but also needs its own unique subsystem that manages these various contact methods. Depending on the size of the customer service organization, an Automatic Call Distribution (ACD) system may be needed as part of the phone system supporting customer service. This system can automatically hold and route calls to the next available representative, and offer advertising or alternate contact methods as part of its "on hold" dialogue. The ACD system also keeps track of caller time on hold, number of calls handled by each representative, and the duration of each call. These metrics can be useful for determining staffing and training requirements.

A similar type of distribution system is needed for automated collection and distribution of e-mail and web-based requests. Moreover, the system supporting the voice-based and data-based requests must somehow coordinate distribution to the Customer Service Representative (CSR) so that there is not contention between the two dissimilar systems. This coordination may be as simple as the CSR going "off-line" on the ACD system while handling these other e-mail or web originated requests.

The ACD generated requests, and potentially the e-mail-based requests, have no way of having their content automatically placed in the requisite databases of the back office systems; therefore this becomes the job of the CSR, who needs a subsystem in which to place the content of the customer conversation.

The customer service subsystem provides a convenient user friendly front end to the other back office systems. This front end provides an input screen to capture customer information and also provides links for retrieving data from other systems. For example, a new service request requires the CSR to collect all the information necessary to initiate service and billing. Information like customer name and address, billing name and address, service plan, the customer equipment's unique identification, and the billing methodology will be necessary to allow the other subsystems to create a new account, authorize service on the customer's equipment, and render the customer a bill on an appropriate interval and in an appropriate manner.

Since this information is part of the customer record, a customer calling to report trouble may only need to provide a name or account number in order for the CSR to get the entire customer record. The reported trouble can be entered into the CSR's computer, which can both update the trouble ticketing system by generating a trouble ticket, as well as query the trouble ticket system and NOCC subsystems to see if there are known system problems that might be the cause of the customer's trouble. If this is the case, then the CSR can inform the customer of the known problem and offer an expected time of resolution.

Design Considerations and Requirements

User requirements

Given that these back office systems provide the man-machine interface to a myriad of network elements, one of the highest priority design considerations is that the systems are "user friendly". After all, this automation is supposed to make the job easier, not more difficult. For those systems that interface directly with a customer (like a self provisioning system), there are additional needs that these systems be secure, error proof, and simple and logical to use. An automated customer support system may be the first interaction a customer has with a company. The impression left by the interaction with the system may color the impression the customer has of the business, so it's important to think through the process and design the interface to make the interaction as simple and pleasant as possible.

The "look and feel" of the user interface is an important area to focus on, since this will provide the user with simple input/output screens necessary to accomplish complex transactions with numerous back office systems. Look and feel can be as simple as color and intuitive layout of the screen or as complex as the logical flow of the information requests and feedback to the user.

The user groups will all have different needs of this user interface. The customer will need the simplest, most logical look and feel because of the infrequency of interaction, while the employees, due to their constant interaction, will require less "handholding." Of course, employee interaction with the back office systems will offer a much greater flexibility to make changes and enter or delete information than is offered to a customer.

Reliability and Security Requirements

These systems form the basis for managing all customer records, billing records, trouble records, and system maintenance records. If they are not functioning, then the business cannot function effectively. Reliability is a key requirement. Reliability comes from both good software design and hardware redundancy for key systems.

Hardware platforms for key systems should be redundant. This redundancy may be provided by splitting the activities over multiple platforms, or by having processor and hardware redundancy as part of the computing platform selected. Further, databases must be provided redundancy via RAID arrays. Also, routine backups should be performed, and at least one recent backup image should be kept off site in a secure location.

Since some of these systems (trouble ticketing, NOCC, and customer self support for example) need to be accessed from and by the "outside world" there is a need to make sure that such open front ends are effectively secured so that no one can get deep enough into the back office systems to cause any breach of security or system problems. Internally focused interfaces like the NOCC and trouble ticketing systems can be secured by requiring VPN and user name and password access. Customer facing systems may not be able to rely on these methods. If this is the case, additional steps must be taken to isolate the system providing customer interface from the rest of the back office hardware and software. This helps to assure that a hacker or other individual with intent to break into the system is stopped before they can cause reach a point where they can cause real damage.

Personnel requirements

Like any IT system, with these systems comes a need for personnel to maintain, operate, and improve the systems. The size and complexity of the systems will in large part determine the personnel needs, but they can be broken down into three basic categories: Hardware maintenance personnel, network maintenance personnel, and software support and development personnel.

Like every other decision made in the business, decisions surrounding back office systems are fraught with trade-offs. Yes, these systems can lead to great efficiencies, but these efficiencies carry a price tag of their own. Not only are there capital costs associated with the hardware and software for these systems, there are ongoing operating expenses like license fees, maintenance and support personnel. Careful consideration of your current and future operating and financial needs is critical in determining which of the available options best fits your unique requirements.

CHAPTER 8

Network Performance Testing and Troubleshooting

- Low RSSI and Noise Levels in a Single Area
- High Noise or Interference Levels
- Throughput Problems Unrelated to Signal or Noise
- Repeater Mode
- Multiple Channel Reuse in a Close Area
- Near/Far Problems
- Hidden Node Problems
- Client Card "Roam"
- Viruses and Trojans
- Troubleshooting Summary

Network Performance Testing and Troubleshooting

RF related network problems manifest themselves in numerous ways. The connection to the network may be unstable, the connection speed may be slow, the user may notice slow response to network queries, or there may be noticeable "holes" in the desired coverage area.

The most common cause of such problems is low SNR. This can be caused by too much path attenuation, by higher than expected interference levels, or may be the result of impaired antenna systems or failing hardware.

Another cause is client "bouncing," which is the tendency for a client at the edge of coverage of two or more base stations to bounce from base station to base station in search of better signal strength. This is the result of low SNR coupled with the search threshold levels set in the client. This problem is quite common in 802.11 systems, but may not be a problem in more complex solutions.

If a baseline signal and noise level was established during the site survey or initial installation, this can be a valuable tool in determining what has changed. If a baseline does not exist, then you need to start out by determining today's baseline, and comparing it to system requirements.

Remember that the network is a bidirectional system. You must understand receive signal strengths and noise levels at both the client and base station. So the first step in troubleshooting is to collect some signal and noise data in the impaired area.

Your survey tool will be useful to rapidly collect information pertaining to signal and noise at the client end of the link. In addition, there is a need to take some number of measurements of the signal and noise seen by the base station. This is most easily done by stopping at a location, opening both the base station monitoring software and the client monitoring software on the PC used to collect data, and collecting all four of the reported numbers for further analysis (client RSSI, client noise, base station RSSI, and base station received noise).

It is critical to have a design baseline that sets the minimum coverage signal strength. That baseline should reflect the actual minimal signal that is necessary for sustainable communication. It will be a derived using the sensitivity of the equipment, the SNR required, the noise floor in the environment and a fade margin. For example, if all clients in an 802.11b network are expected to maintain an 11 Mbps connection, then the required RSSI will be –83 dBm if the noise floor is –100 dBm. In addition, at least 10 dB of fade margin will be necessary in order to compensate for multipath effects, thus helping to assure the signal remains above the required minimum signal threshold. Therefore design coverage criteria should be –73 dBm in all locations where the system will be used. If the expected throughput is reduced to 1 Mbps, then design criteria can be reduced to – 83 dBm, assuming the system supports 1 Mbps connections at an RSSI of –93 dBm.

If system noise floor is above –100 dBm, then additional signal will be required in order to maintain SNR. For every 1 dB of additional noise, the required signal strength will increase by 1 dB. In the case of the previous example, if the noise floor were measured at –90 dBm instead of –100 dBm, then the receive signal strength necessary to support a 11 Mbps connection would increase to –63 dBm.

Once the data is collected it can be analyzed. The measurement results will be used to determine a set of scenarios, which are reflections of the data collected. Each scenario is specific in the nature of the fault, and normally has a limited number of causes and possible solutions. Starting with RSSI related issues only, then moving into more complex interference and noise related issues, the breakdown below offers some analysis of the differing failure modes.

Low RSSI and Noise Levels in a Single Area

Low RSSI on all base stations in both directions in all locations:

Likely causes:

- Defective client

- Bad measurement technique

- Defective test equipment

Resolution: Evaluate test equipment and procedures. Retest in a known environment in order to assure that measurements are accurate. Once testing is showing accurate results, begin analysis again.

Low RSSI at all base stations in all locations:

Likely causes:

- Defective client

- Low power client device not providing a balanced path

Resolution: Replace client and retest. If problem persists try a known good client device.

Low RSSI on a single base station in both directions in all locations:

Likely causes:

- Antenna system failure

- Defective antenna

- Misdirected antenna

- Insufficient downtilt (especially on Omni antennas)

- Defective connector or feedline

- Water in feedline or connector due to improper sealing

Resolution: Inspect affected base station's feedline and antenna system. Assure that the antenna main beam is accurately pointed and that downtilt is appropriate for the area to be covered. Try replacing the antenna and feedline with a known good substitute.

Low RSSI on a single base station on base station direction only, all locations:

Likely causes:

- Defective base station receiver

Resolution: Replace base station

Low RSSI on a single base station on client direction only, all locations:

Likely causes:

- Defective base station transmitter

Resolution: Replace base station

Low RSSI in both directions in certain locations but OK in others:

Likely causes:

- Local path attenuation is greater than expected in the impaired location.

Resolution: Move base station to a location that has better coverage of affected area. Add base station to cover affected area.

High Noise or Interference Levels

High noise in both directions in all locations:

Likely causes:

- Bad measurement technique
- Defective test equipment
- High level of noise from other source

Resolution: Evaluate test equipment and procedures. Retest in a known environment in order to assure that measurements are accurate. Once testing is showing accurate results, begin analysis again. If noise persists, use a directional antenna in conjunction with the test equipment in order to attempt to find the noise source. In the case of 802.11a or 802.11b, or any other technology using unlicensed spectrum, you have no standing to cause an interferer to cease operation. Luckily most interferers are low bandwidth devices and should cause noise on single channels rather than across the entire band. Changing your operating channel to one with less interference should resolve the problem.

High noise levels at all base stations in all locations:

Likely causes: The base station, by nature of its location and antenna, will see more noise than the client in a single semi shielded location will see. It is common to see the base station report a noise level 3 to 10 dB higher than the client. This is not necessarily a problem, and associated client SNR should be analyzed to assure there is sufficient signal to overcome the noise. For example, all 802.11b devices have a receive sensitivity which is based on a noise floor of –100 dBm. An 802.11 base station that requires –93 dBm of signal to maintain a 1 Mbps connection, only requires –93 dBm when the noise floor is –100 dBm. If the measured noise floor is –90, then the required signal rises by 10 dB also, so the base station will need –83 dBm of signal to maintain the connection.

Resolution: Assure that signal is sufficient to maintain adequate SNR.

If there are multiple base stations operating on common channels in the area, they or their associated clients may be contributing to the noise floor. If unlicensed bands are being used, another user's equipment may be contributing to the noise. If this is the case, the channelization of the system or the reuse plan may need to be modified.

Resolution: First, determine whether the interference is being generated by equipment associated with your network. Do this by testing at a noisy location, while turning off other co-channel base stations, and see how the noise floor is affected. If significant improvement in noise is noted, then the reuse plan is ineffective and will need to be modified. If no improvement is noted, the noise is coming from an outside source. Channel changes may be an effective way of reducing the noise seen by individual base stations.

High noise on a single base station in both directions in all locations:

Likely causes: This is an unlikely event. If it exists, look for an on channel interferer, either another interfering base station, or a noise source like a cordless phone system, baby monitor, wireless video system, and so forth. If using unlicensed spectrum, or if operating in licensed spectrum, use a spectrum analyzer to look for noise sources like Intermod or transmitters in use in the areas that have failed and are generating out-of-band energy.

Resolution: Change channel of affected base station. This may have a ripple effect on the reuse plan in the area.

High noise on a single base station in base station direction only, all locations:

Likely causes:

- Defective base station receiver

- Noise source in field of view of base station antenna

Resolution: Change channel, disconnect antenna and see if the base station still reports high noise. If so the base station may be defective, replace base station. Move base station and antenna to a location where noise is not seen, add base station in location where noise is not seen.

High noise on a single base station in client direction only, all locations:

Likely causes:

- Defective base station transmitter

Resolution: Replace or repair base station

High noise in both directions in certain locations but OK in others:

Likely causes:

- Local noise source such as discussed above.

Resolution: Add base station to cover affected area. Change existing base station channel to a clean one.

High noise in client direction in certain areas:

Likely causes:

- Localized noise source such as discussed above

Resolution: Add base station to affected area, change existing base station channel to a clean one.

Throughput Problems Unrelated to Signal or Noise

In addition to noise or signal related issues, certain deployments can cause reduced throughput due to the nature of the deployment leading to channel usage contention among the various devices.

Repeater Mode

Operating 802.11 base stations as repeaters is useful in areas where the repeaters can be shielded from each other. In a deployment where the repeaters can see each other's signals, system throughput will be decreased. This is because of the way 802.11b manages bandwidth allocation. The standard uses CSMA/CA or carrier sense multiple access / collision avoidance. This simply means that multiple users are supported because 802.11 allows only one at a time to use the channel. Everyone listens to the channel before transmitting. If the channel is unused, then the user can transmit. If, on the other hand, a signal is heard, then the user will "back off" by a random time interval and try again after that interval has expired.

In a repeated network where the repeaters can see each other, this leads to contention problems for the repeaters and the control station. Let's take an example of a master station and four repeaters. The master sends its packets to all repeaters. The repeaters hear and accept the packets. Now the fastest repeater accesses the channel to rebroadcast the packets. While this repeater is transmitting, all other repeaters are in a back-off mode. Eventually, they will all get a chance to repeat the message. Unfortunately, the result of this is that the packet takes up the channel five times longer than it would have taken if no repeats were used, thus throughput is reduced five-fold. Worse than this is the fact that the packet belonged to a single user, so most of the repeater rebroadcasts were a waste of time.

This channel contention causes another problem for the user: latency and lost packets. Let's assume that the client waiting for the packet is covered by the first repeater. He may immediately get the packet and try to send a response. Because he probably can not hear the other repeaters, his response may come at a time when another repeater is broadcasting. Since the channel is in use and the other repeaters are being heard by the first repeater, there is a good chance that the client's packet will be lost due to interference. This leads to a need to resend packets, thus further reducing network throughput.

In a situation where repeaters see each other, repeaters should be avoided. Instead use equipment that offers dual radio capability, so that one radio can be set as a client and the other as a base station rebroadcasting on a different channel.

Multiple Channel Reuse in a Close Area

CSMA/CA leads to similar problems in areas where multiple base stations are sharing a common channel. Again the base stations may be hearing each other but the clients because of their shielded location cannot hear all base stations. This causes interference to develop at the base station because the base station is hearing multiple simultaneous transmissions, and cannot discriminate between the desired and undesired signals. The result is packet loss and retransmissions.

This situation can be identified by using the client software to identify signal strength by MAC address outside in the area near each base station. If the signal from co-channel base stations can be seen here and not in the area where the users are located, then there will be a problem.

Better reuse planning, power reduction, or implementing RTS/CTS in an 802.11-based network may help the situation.

Near/Far Problems

Another issue with certain radio-based systems like 802.11 is known as the near/far problem. In this situation stronger signals mask weaker ones, leading to a disproportionate amount of bandwidth being allocated to users that consistently present a stronger signal to the base station (for example, those closest to the base station). RTS/CTS can help this situation, but will not completely eliminate it. Using higher power client cards or external high gain antennas in weak areas may also help.

Hidden Node Problems

This is a problem experienced by 802.11 or other CSMA-based networks. It occurs when client stations can hear a common AP but not each other, which is a common problem in campus or community coverage networks. Because CSMA shares a channel by listening to assure a channel is clear, if stations communicating to a common base station cannot hear each other they cannot effectively share the channel. This leads to the opportunity for multiple stations to assume the channel is clear, and begin simultaneously transmitting. The result is that the base station hears multiple simultaneous signals. If one is significantly stronger, it may capture the channel and get through, while the others are lost. If the signals are all of equivalent strength, then all will be lost.

RTS/CTS can help this situation, but will not completely eliminate it. If it is financially viable, changing to a protocol that is not dependent on CSMA/CA is a better long term solution to this problem.

Client Card "Roam"

Some standards-based and proprietary solutions are designed to support "roaming" or handoff between base stations or APs. This allows users to move about an area and shift from base station to base station based upon the strongest signal. Unfortunately, the default setting in many cards forces the card always to look for a better signal. As soon as the signal fades, the card tries to reassociate with another base station, often losing connectivity and packets in the process. Because of the nature of the fading environment, this "new best signal" may not be the best for long, and the card will again try to reassociate. Early Lucent drivers and Cisco 802.11b drivers had the ability to select what they called *base station density* to low, medium, or high. All clients should be set appropriate to the density of the APs in the network. This tells a client that there are many base station's and to hold onto the one it has until it drops to unacceptable RSSI and stays there. This will help to stabilize associations between base stations and clients and reduce lost packets and unstable connections.

Viruses and Trojans

We live in a world where Internet and email borne viruses and Trojan programs are proliferating. Worse yet, many computer users are either ignorant of the situation, or choose to ignore it.

Unfortunately, such programs can cause problems to wireless network throughput. Because many of these programs actively seek to infect others or to send some information to another computer, they utilize network bandwidth. If many computers on a wireless network become infected, the viruses begin using a considerable amount of bandwidth. This bandwidth is no longer available to carry legitimate traffic.

This situation may require the use of an Ethernet packet analyzer to allow you to inspect the traffic on the wireless node and see if a large proportion of it is virus or Trojan related. Ultimately, the only way to solve this problem is to have customers run anti virus software which is kept updated, or to identify users with viruses and disconnect them from service until the virus problem in their computer is eliminated.

Troubleshooting Summary

Every system is unique, so there is no way for this book to cover all possible troubles you may encounter in a network. The problems discussed are merely a helpful identification of issues commonly found in an RF system. The examples most certainly do not identify every source of trouble you are likely to find, however they are useful as a starting point.

Like any other troubleshooting process, the RF system troubleshooting process requires a systematic approach, an intimate understanding of the variables in play in your network, and an understanding of RF and the environmental factors that affect it. Having these basics will allow you to approach the problem logically, define its nature, and determine from what possible sources the trouble arises. Eliminating those issues that cannot be the source of the problem will allow you to hone in on the problem by focusing on the most probable causes and then eliminating all but the root cause of the problem.

About the Author

Ron Olexa has been actively involved in the design, deployment, and operation of wireless communications systems for over 30 years. He has designed systems as simple as individual point-to-point links and as complex as national scale GSM networks. In the early 1980's, Ron worked in a senior management role with a number of the LIN Broadcasting and Metromedia cellular markets, where he was responsible for the design, deployment, and operation of some of the first cellular systems deployed in the top five U.S. markets. In the late 1980's and early 1990's, he was responsible for the initial design of a number of international GSM and CDMA system designs for PacTel Cellular (later known as Airtouch and now part of Vodaphone). The mid-1990's brought a new technical challenge, and Ron joined DialCall Inc., where he assembled a team that designed, constructed and managed a network based upon Motorola's iDEN technology, the same technology currently used by Nextel Communications.

In the late 1990's, Ron shifted from wireless voice networks to the emerging wireless data industry. He was responsible for system designs of point-to-point and point-to-multipoint data networks using millimeter microwave spectrum. From 2000 to present, Ron has run a consulting company that has provided technical support and business planning guidance to projects as diverse as satellite communications systems and 802.11 hotspot and hotzone implementations.

Ron can be reached through his company website: www.wirelessimplementation.com.

List of Acronyms

AC	Alternating Current
AM	Amplitude Modulation
AP	Access Point
APC	Automatic Power Control
ATM	Asynchronous Transfer Mode
BS	Base Station
BPSK	BiPhase Shift Keying
CAPEX	Capital Expense
CAT5	Category 5 Cable
CC&R	Conditions, Covenants, and Restrictions
CCK	Complimentary Code Keying
C/I	Carrier to Interference Ratio
C/I+N	Carrier-to-Noise plus Interference ratio
C/N	Carrier-to-Noise Ratio
CPE	Customer Premise Equipment
CSMA/CA	Carrier Sense Multiple Access / Collision Avoidance
CSMA/CD	Carrier Sense Multiple Access / Collision Detect
CTS	Clear to Send
CW	Continuous Wave
dB	Decibel
dBd	Decibel Gain Referenced to a Dipole Antenna

dBi	Decibel gain referenced to an Isotropic Antenna
dBm	Decibels referenced to 1 milliwatt
dBu	Decibels referenced to 1 microvolt
dBW	Decibels referenced to 1 Watt
DC	Direct Current
DHCP	Dynamic Host Configuration Protocol
DPC	Dynamic Power Control
DSL	Digital Subscriber Line
DSSS	Direct Sequence Spread Spectrum
Eb/No	Energy per Bit to Noise Ratio
EM	Electro Magnetic
FCC	Federal Communications Commission
FDD	Frequency Division Duplexing
FDM	Frequency Division Multiplexing
FDMA	Frequency Division Multiple Access
FHSS	Frequency Hopping Spread Spectrum
FM	Frequency Modulation
FSL	Free Space Loss
GBPS	Gigabits Per Second
GPS	Global Positioning System
GUI	Graphical User Interface
Hz	Hertz or Cycles per Second
IEEE	Institute of Electric and Electronic Engineers
IP	Internet Protocol
ISP	Internet Service Provider

LAN	Local Area Network
LO	Local Oscillator
LOS	Line of Sight
MAC	Media Access Layer
MAN	Metropolitan Area Network
MBPS	Megabits per Second
MDU	Multiple Dwelling Unit
Modem	Modulator/Demodulator
NAT	Network Address Translation
NEC	National Electrical Code
NLOS	Non or Near Line of Sight
NOCC	Network Operations and Control Center
NPRM	Notice of Proposed Rule Making
OFDM	Orthogonal Frequency Division Multiplexing
OPEX	Operating Expense
PCS	Personal Communication Service
PHY	Physical Layer
PM	Phase Modulation
PMP	Point-to-Multipoint
PtMP	Point-to-Multipoint
PTP	Point-to-Point
QAM	Quadrature Amplitude Modulation or Quaternary Amplitude Modulation
QOS	Quality of Service
QPSK	Quaternary Phase Shift Keying

RADIUS	Remote Authentication Dial In Server
RF	Radio Frequency
RSSI	Relative Signal Strength Indicator
RTS	Request to Send
SNR	Signal-to-Noise Ratio
SSID	Service Set Identifier
TDD	Time Division Duplex
TDM	Time Division Multiplexing
TDMA	Time Division Multiple Access
USGS	United States Geological Survey
UWB	Ultra Wideband
VAR	Value Added Reseller
VOIP	Voice over IP
Wi-Fi	Wireless Fidelity
WiMAX	Worldwide Interoperability for Microwave Access
WISP	Wireless ISP

Index

Numbers

100Base-FX, 155
100Base-T, 166
10Base-FL, 155
802.11b, xiii
802.11x, 6
802.16, 6
802.20, 6

A

Amateur Radio operators, xiv
amplitude modulation, 44
antennas, 35, 36
    antenna downtilt, 125
    antenna gain, 79
    antenna height, 57
    antenna selection, 57, 125
ATM, 15
attenuation, 136
automatic call distribution (ACD), 207
average usage, 145

B

backhaul, 181
back office systems, 201
Barker code, 61
    Barker code direct sequence spread, 10
billing data collection, 203
billing system, 204
BiPhase Shift Keying, 49
BPSK, 10, 50

C

C/I ratio, 159
cable, 79
capacitive reactance, 32
capacity, 159
capacity, 132
CAPEX, xvi, xvii, 159
carrier sense multiple access/collision
    avoidance (CSMA/CA), 10
carrier to noise or interference ratio
    (C/I, C/N, C/I+N), 48
CAT5, 150
channelization, 46
channel reuse, 220
Class A, AB, B, C, and D, 35
client hardware, 8
coax, 35
code division multiple access, 60
collisions, 148
common reuse patterns, 89
complexity, 159
complimentary code keying (CCK), 10, 61
comprehensive site survey, 106
connector, 79
cost, 132, 159
coverage, 132, 159
customer authorization system, 202
customer premise equipment (CPE), 113

ELSEVIER SCIENCE CD-ROM LICENSE AGREEMENT

PLEASE READ THE FOLLOWING AGREEMENT CAREFULLY BEFORE USING THIS CD-ROM PRODUCT. THIS CD-ROM PRODUCT IS LICENSED UNDER THE TERMS CONTAINED IN THIS CD-ROM LICENSE AGREEMENT ("Agreement"). BY USING THIS CD-ROM PRODUCT, YOU, AN INDIVIDUAL OR ENTITY INCLUDING EMPLOYEES, AGENTS AND REPRESENTATIVES ("You" or "Your"), ACKNOWLEDGE THAT YOU HAVE READ THIS AGREEMENT, THAT YOU UNDERSTAND IT, AND THAT YOU AGREE TO BE BOUND BY THE TERMS AND CONDITIONS OF THIS AGREEMENT. ELSEVIER SCIENCE INC. ("Elsevier Science") EXPRESSLY DOES NOT AGREE TO LICENSE THIS CD-ROM PRODUCT TO YOU UNLESS YOU ASSENT TO THIS AGREEMENT. IF YOU DO NOT AGREE WITH ANY OF THE FOLLOWING TERMS, YOU MAY, WITHIN THIRTY (30) DAYS AFTER YOUR RECEIPT OF THIS CD-ROM PRODUCT RETURN THE UNUSED CD-ROM PRODUCT AND ALL ACCOMPANYING DOCUMENTATION TO ELSEVIER SCIENCE FOR A FULL REFUND.

DEFINITIONS

As used in this Agreement, these terms shall have the following meanings:

"Proprietary Material" means the valuable and proprietary information content of this CD-ROM Product including all indexes and graphic materials and software used to access, index, search and retrieve the information content from this CD-ROM Product developed or licensed by Elsevier Science and/or its affiliates, suppliers and licensors.

"CD-ROM Product" means the copy of the Proprietary Material and any other material delivered on CD-ROM and any other human-readable or machine-readable materials enclosed with this Agreement, including without limitation documentation relating to the same.

OWNERSHIP

This CD-ROM Product has been supplied by and is proprietary to Elsevier Science and/or its affiliates, suppliers and licensors. The copyright in the CD-ROM Product belongs to Elsevier Science and/or its affiliates, suppliers and licensors and is protected by the national and state copyright, trademark, trade secret and other intellectual property laws of the United States and international treaty provisions, including without limitation the Universal Copyright Convention and the Berne Copyright Convention. You have no ownership rights in this CD-ROM Product. Except as expressly set forth herein, no part of this CD-ROM Product, including without limitation the Proprietary Material, may be modified, copied or distributed in hardcopy or machine-readable form without prior written consent from Elsevier Science. All rights not expressly granted to You herein are expressly reserved. Any other use of this CD-ROM Product by any person or entity is strictly prohibited and a violation of this Agreement.

SCOPE OF RIGHTS LICENSED (PERMITTED USES)

Elsevier Science is granting to You a limited, non-exclusive, non-transferable license to use this CD-ROM Product in accordance with the terms of this Agreement. You may use or provide access to this CD-ROM Product on a single computer or terminal physically located at Your premises and in a secure network or move this CD-ROM Product to and use it on another single computer or terminal at the same location for personal use only, but under no circumstances may You use or provide access to any part or parts of this CD-ROM Product on more than one computer or terminal simultaneously.

You shall not (a) copy, download, or otherwise reproduce the CD-ROM Product in any medium, including, without limitation, online transmissions, local area networks, wide area networks, intranets, extranets and the Internet, or in any way, in whole or in part, except that You may print or download limited portions of the Proprietary Material that are the results of discrete searches; (b) alter, modify, or adapt the CD-ROM Product, including but not limited to decompiling, disassembling, reverse engineering, or creating derivative works, without the prior written approval of Elsevier Science; (c) sell, license or otherwise distribute to third parties the CD-ROM Product or any part or parts thereof; or (d) alter, remove, obscure or obstruct the display of any copyright, trademark or other proprietary notice on or in the CD-ROM Product or on any printout or download of portions of the Proprietary Materials.

RESTRICTIONS ON TRANSFER

This License is personal to You, and neither Your rights hereunder nor the tangible embodiments of this CD-ROM Product, including without limitation the Proprietary Material, may be sold, assigned, transferred or sub-licensed to any other person, including without limitation by operation of law, without the prior written consent of Elsevier Science. Any purported sale, assignment, transfer or sublicense without the prior written consent of Elsevier Science will be void and will automatically terminate the License granted hereunder.

TERM

This Agreement will remain in effect until terminated pursuant to the terms of this Agreement. You may terminate this Agreement at any time by removing from Your system and destroying the CD-ROM Product. Unauthorized copying of the CD-ROM Product, including without limitation, the Proprietary Material and documentation, or otherwise failing to comply with the terms and conditions of this Agreement shall result in automatic termination of this license and will make available to Elsevier Science legal remedies. Upon termination of this Agreement, the license granted herein will terminate and You must immediately destroy the CD-ROM Product and accompanying documentation. All provisions relating to proprietary rights shall survive termination of this Agreement.

LIMITED WARRANTY AND LIMITATION OF LIABILITY

NEITHER ELSEVIER SCIENCE NOR ITS LICENSORS REPRESENT OR WARRANT THAT THE INFORMATION CONTAINED IN THE PROPRIETARY MATERIALS IS COMPLETE OR FREE FROM ERROR, AND NEITHER AS-SUMES, AND BOTH EXPRESSLY DISCLAIM, ANY LIABILITY TO ANY PERSON FOR ANY LOSS OR DAMAGE CAUSED BY ERRORS OR OMISSIONS IN THE PROPRIETARY MATERIAL, WHETHER SUCH ERRORS OR OMIS-SIONS RESULT FROM NEGLIGENCE, ACCIDENT, OR ANY OTHER CAUSE. IN ADDITION, NEITHER ELSEVIER SCIENCE NOR ITS LICENSORS MAKE ANY REPRESENTATIONS OR WARRANTIES, EITHER EXPRESS OR IMPLIED, REGARDING THE PERFORMANCE OF YOUR NETWORK OR COMPUTER SYSTEM WHEN USED IN CONJUNCTION WITH THE CD-ROM PRODUCT.

If this CD-ROM Product is defective, Elsevier Science will replace it at no charge if the defective CD-ROM Product is returned to Elsevier Science within sixty (60) days (or the greatest period allowable by applicable law) from the date of shipment.

Elsevier Science warrants that the software embodied in this CD-ROM Product will perform in substantial compliance with the documentation supplied in this CD-ROM Product. If You report significant defect in performance in writing to Elsevier Science, and Elsevier Science is not able to correct same within sixty (60) days after its receipt of Your notification, You may return this CD-ROM Product, including all copies and documentation, to Elsevier Science and Elsevier Science will refund Your money.

YOU UNDERSTAND THAT, EXCEPT FOR THE 60-DAY LIMITED WARRANTY RECITED ABOVE, ELSEVIER SCIENCE, ITS AFFILIATES, LICENSORS, SUPPLIERS AND AGENTS, MAKE NO WARRANTIES, EXPRESSED OR IMPLIED, WITH RESPECT TO THE CD-ROM PRODUCT, INCLUDING, WITHOUT LIMITATION THE PROPRIETARY MATERIAL, AN SPECIFICALLY DISCLAIM ANY WARRANTY OF MERCHANTABILITY OR FITNESS FOR A PARTICULAR PURPOSE.

If the information provided on this CD-ROM contains medical or health sciences information, it is intended for professional use within the medical field. Information about medical treatment or drug dosages is intended strictly for professional use, and because of rapid advances in the medical sciences, independent verification f diagnosis and drug dosages should be made.

IN NO EVENT WILL ELSEVIER SCIENCE, ITS AFFILIATES, LICENSORS, SUPPLIERS OR AGENTS, BE LIABLE TO YOU FOR ANY DAMAGES, INCLUDING, WITHOUT LIMITATION, ANY LOST PROFITS, LOST SAVINGS OR OTHER INCIDENTAL OR CONSEQUENTIAL DAMAGES, ARISING OUT OF YOUR USE OR INABILITY TO USE THE CD-ROM PRODUCT REGARDLESS OF WHETHER SUCH DAMAGES ARE FORESEEABLE OR WHETHER SUCH DAMAGES ARE DEEMED TO RESULT FROM THE FAILURE OR INADEQUACY OF ANY EXCLUSIVE OR OTHER REMEDY.

U.S. GOVERNMENT RESTRICTED RIGHTS

The CD-ROM Product and documentation are provided with restricted rights. Use, duplication or disclosure by the U.S. Government is subject to restrictions as set forth in subparagraphs (a) through (d) of the Commercial Computer Restricted Rights clause at FAR 52.22719 or in subparagraph (c)(1)(ii) of the Rights in Technical Data and Computer Software clause at DFARS 252.2277013, or at 252.2117015, as applicable. Contractor/Manufacturer is Elsevier Science Inc., 655 Avenue of the Americas, New York, NY 10010-5107 USA.

GOVERNING LAW

This Agreement shall be governed by the laws of the State of New York, USA. In any dispute arising out of this Agreement, you and Elsevier Science each consent to the exclusive personal jurisdiction and venue in the state and federal courts within New York County, New York, USA.